T0331620

A Student's Guide to Geophysical Equations

The advent of accessible student computing packages has meant that geophysics students can now easily manipulate datasets and gain first-hand modeling experience – essential in developing an intuitive understanding of the physics of the Earth. Yet to gain a more in-depth understanding of the physical theory, and to be able to develop new models and solutions, it is necessary to be able to derive the relevant equations from first principles.

This compact, handy book fills a gap left by most modern geophysics textbooks, which generally do not have space to derive all of the important formulae, showing the intermediate steps. This guide presents full derivations for the classical equations of gravitation, gravity, tides, Earth rotation, heat, geomagnetism, and foundational seismology, illustrated with simple schematic diagrams. It supports students through the successive steps and explains the logical sequence of a derivation – facilitating self-study and helping students to tackle homework exercises and prepare for exams.

WILLIAM LOWRIE was born in Hawick, Scotland, and attended the University of Edinburgh, where he graduated in 1960 with first-class honors in physics. He achieved a masters degree in geophysics at the University of Toronto and, in 1967, a doctorate at the University of Pittsburgh. After two years in the research laboratory of Gulf Oil Company he became a researcher at the Lamont-Doherty Geological Observatory of Columbia University. In 1974 he was elected professor of geophysics at the ETH Zürich (Swiss Federal Institute of Technology in Zurich), Switzerland, where he taught and researched until retirement in 2004. His research in rock magnetism and paleomagnetism consisted of deducing the Earth's magnetic field in the geological past from the magnetizations of dated rocks. The results were applied to the solution of geologic-tectonic problems, and to analysis of the polarity history of the geomagnetic field. Professor Lowrie has authored 135 scientific articles and a second edition of his acclaimed 1997 textbook *Fundamentals of Geophysics* was published in 2007. He has been President of the European Union of Geosciences (1987–9) and Section President and Council member of the American Geophysical Union (2000–2). He is a Fellow of the American Geophysical Union and a Member of the Academia Europaea.

A Student's Guide
to Geophysical Equations

WILLIAM LOWRIE
Institute of Geophysics
Swiss Federal Institute of Technology
Zurich, Switzerland

CAMBRIDGE
UNIVERSITY PRESS

CAMBRIDGE
UNIVERSITY PRESS

University Printing House, Cambridge CB2 8BS, United Kingdom

Cambridge University Press is part of the University of Cambridge.

It furthers the University's mission by disseminating knowledge in the pursuit of education, learning and research at the highest international levels of excellence.

www.cambridge.org
Information on this title: www.cambridge.org/9781107005846

First published 2011
4th printing 2014

A catalogue record for this publication is available from the British Library

Library of Congress Cataloguing in Publication data
Lowrie, William, 1939–
A student's guide to geophysical equations / William Lowrie.
p. cm.
Includes bibliographical references and index.
ISBN 978-1-107-00584-6 (hardback)
1. Geophysics – Mathematics – Handbooks, manuals, etc.
2. Physics – Formulae – Handbooks, manuals, etc. 3. Earth – Handbooks, manuals, etc.
I. Title.
QC809.M37L69 2011
550.1'51525–dc22
2011007352

ISBN 978-1-107-00584-6 Hardback
ISBN 978-0-521-18377-2 Paperback

This book is dedicated to Marcia

Contents

Preface

This work was written as a supplementary text to help students understand the mathematical steps in deriving important equations in classical geophysics. It is not intended to be a primary textbook, nor is it intended to be an introduction to modern research in any of the topics it covers. It originated in a set of handouts, a kind of "do-it-yourself" manual, that accompanied a course I taught on theoretical geophysics. The lecture aids were necessary for two reasons. First, my lectures were given in German and there were no comprehensive up-to-date texts in the language; the recommended texts were in English, so the students frequently needed clarification. Secondly, it was often necessary to explain classical theory in more detail than one finds in a multi-topic advanced textbook. To keep such a book as succinct as possible, the intermediate steps in the mathematical derivation of a formula must often be omitted. Sometimes the unassisted student cannot fill in the missing steps without individual tutorial assistance, which is usually in short supply at most universities, especially at large institutions. To help my students in these situations, the "do-it-yourself" text that accompanied my lectures explained missing details in the derivations. This is the background against which I prepared the present guide to geophysical equations, in the hope that it might be helpful to other students at this level of study.

The classes that I taught to senior grades were largely related to potential theory and primarily covered topics other than seismology, since this was the domain of my colleagues and better taught by a true seismologist than by a paleomagnetist! Theoretical seismology is a large topic that merits its own treatment at an advanced level, and there are several textbooks of classical and modern vintage that deal with this. However, a short chapter on the relationship of stress, strain, and the propagation of seismic waves is included here as an introduction to the topic.

Computer technology is an essential ingredient of progress in modern geophysics, but a well-trained aspiring geophysicist must be able to do more than apply advanced software packages. A fundamental mathematical understanding is needed in order to formulate a geophysical problem, and numerical computational skills are needed to solve it. The techniques that enabled scientists to understand much about the Earth in the pre-computer era also underlie much of modern methodology. For this reason, a university training in geophysics still requires the student to work through basic theory. This guide is intended as a companion in that process.

Historically, most geophysicists came from the field of physics, for which geophysics was an applied science. They generally had a sound training in mathematics. The modern geophysics student is more likely to have begun studies in an Earth science discipline, the mathematical background might be heavily oriented to the use of tailor-made packaged software, and some students may be less able to handle advanced mathematical topics without help or tutoring. To fill these needs, the opening chapter of this book provides a summary of the mathematical background for topics handled in subsequent chapters.

Acknowledgments

In writing this book I have benefited from the help and support of various people. At an early stage, anonymous proposal reviewers gave me useful suggestions, not all of which have been acted on, but all of which were appreciated. Each chapter was read and checked by an obliging colleague. I wish to thank Dave Chapman, Rob Coe, Ramon Egli, Chris Finlay, Valentin Gischig, Klaus Holliger, Edi Kissling, Emile Klingelé, Alexei Kuvshinov, Germán Rubino, Rolf Sidler, and Doug Smylie for their corrections and suggestions for improvement. The responsibility for any errors that escaped scrutiny is, of course, mine. I am very grateful to Derrick Hasterok and Dave Chapman for providing me with an unpublished figure from Derrick's Ph.D. thesis. Dr. Susan Francis, Senior Commissioning Editor at Cambridge University Press, gave me constant support and friendly encouragement throughout the many months of writing, for which I am sincerely grateful. Above all, I thank my wife Marcia for her generous tolerance of the intrusion of this project into our retirement activities.

1

Mathematical background

1.1 Cartesian and spherical coordinates

Two systems of orthogonal coordinates are used in this book, sometimes interchangeably. Cartesian coordinates (x, y, z) are used for a system with rectangular geometry, and spherical polar coordinates (r, θ, ϕ) are used for spherical geometry. The relationship between these reference systems is shown in Fig. 1.1(a). The convention used here for spherical geometry is defined as follows: the radial distance from the origin of the coordinates is denoted r, the polar angle θ (geographic equivalent: the co-latitude) lies between the radius and the z-axis (geographic equivalent: Earth's rotation axis), and the azimuthal angle ϕ in the x–y plane is measured from the x-axis (geographic equivalent: longitude). Position on the surface of a sphere (constant r) is described by the two angles θ and ϕ. The Cartesian and spherical polar coordinates are linked as illustrated in Fig. 1.1(b) by the relationships

$$
\begin{aligned}
x &= r \sin \theta \cos \phi \\
y &= r \sin \theta \sin \phi \\
z &= r \cos \theta
\end{aligned}
\tag{1.1}
$$

1.2 Complex numbers

The numbers we most commonly use in daily life are *real* numbers. Some of them are also *rational* numbers. This means that they can be expressed as the quotient of two integers, with the condition that the denominator of the quotient must not equal zero. When the denominator is 1, the real number is an integer. Thus 4, 4/5, 123/456 are all rational numbers. A real number can also be *irrational*, which means it cannot be expressed as the quotient of two integers.

1

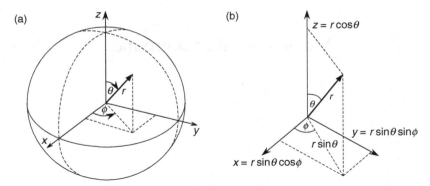

Fig. 1.1. (a) Cartesian and spherical polar reference systems. (b) Relationships between the Cartesian and spherical polar coordinates.

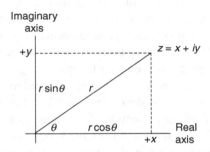

Fig. 1.2. Representation of a complex number on an Argand diagram.

Familiar examples are π, e (the base of natural logarithms), and some square roots, such as $\sqrt{2}$, $\sqrt{3}$, $\sqrt{5}$, etc. The irrational numbers are real numbers that do not terminate or repeat when expressed as decimals.

In certain analyses, such as determining the roots of an equation, it is necessary to find the square root of a negative real number, e.g. $\sqrt{(-y^2)}$, where y is real. The result is an *imaginary* number. The negative real number can be written as $(-1)y^2$, and its square root is then $\sqrt{(-1)}y$. The quantity $\sqrt{(-1)}$ is written i and is known as the imaginary unit, so that $\sqrt{(-y^2)}$ becomes $\pm iy$.

A complex number comprises a real part and an imaginary part. For example, $z = x + iy$, in which x and y are both real numbers, is a complex number with a *real part x* and an *imaginary part y*. The composition of a complex number can be illustrated graphically with the aid of the *complex plane* (Fig. 1.2). The real part is plotted on the horizontal axis, and the imaginary part on the vertical axis. The two independent parts are orthogonal on the plot and the complex number z

is represented by their vector sum, defining a point on the plane. The distance r of the point from the origin is given by

$$r = \sqrt{x^2 + y^2} \tag{1.2}$$

The line joining the point to the origin makes an angle θ with the real (x-)axis, and so r has real and imaginary components $r \cos \theta$ and $r \sin \theta$, respectively. The complex number z can be written in polar form as

$$z = r(\cos \theta + i \sin \theta) \tag{1.3}$$

It is often useful to write a complex number in the exponential form introduced by Leonhard Euler in the late eighteenth century. To illustrate this we make use of infinite power series; this topic is described in Section 1.10. The exponential function, exp(x), of a variable x can be expressed as a power series as in (1.135). On substituting $x = i\theta$, the power series becomes

$$\begin{aligned}
\exp(i\theta) &= 1 + (i\theta) + \frac{(i\theta)^2}{2!} + \frac{(i\theta)^3}{3!} + \frac{(i\theta)^4}{4!} + \frac{(i\theta)^5}{5!} + \frac{(i\theta)^6}{6!} + \cdots \\
&= 1 + \frac{(i\theta)^2}{2!} + \frac{(i\theta)^4}{4!} + \frac{(i\theta)^6}{6!} \cdots + (i\theta) + \frac{(i\theta)^3}{3!} + \frac{(i\theta)^5}{5!} + \cdots \\
&= \left(1 - \frac{\theta^2}{2!} + \frac{\theta^4}{4!} - \frac{\theta^6}{6!} + \cdots\right) + i\left(\theta - \frac{\theta^3}{3!} + \frac{\theta^5}{5!} + \cdots\right)
\end{aligned} \tag{1.4}$$

Comparison with (1.135) shows that the first bracketed expression on the right is the power series for $\cos \theta$; the second is the power series for $\sin \theta$. Therefore

$$\exp(i\theta) = \cos \theta + i \sin \theta \tag{1.5}$$

On inserting (1.5) into (1.3), the complex number z can be written in exponential form as

$$z = r \exp(i\theta) \tag{1.6}$$

The quantity r is the *modulus* of the complex number and θ is its *phase*.

Conversely, using (1.5) the cosine and sine functions can be defined as the sum or difference of the complex exponentials $\exp(i\theta)$ and $\exp(-i\theta)$:

$$\begin{aligned}
\cos \theta &= \frac{\exp(i\theta) + \exp(-i\theta)}{2} \\
\sin \theta &= \frac{\exp(i\theta) - \exp(-i\theta)}{2i}
\end{aligned} \tag{1.7}$$

1.3 Vector relationships

A scalar quantity is characterized only by its magnitude; a vector has both magnitude and direction; a unit vector has unit magnitude and the direction of the quantity it represents. In this overview the unit vectors for Cartesian coordinates (x, y, z) are written $(\mathbf{e}_x, \mathbf{e}_y, \mathbf{e}_z)$; unit vectors in spherical polar coordinates (r, θ, ϕ) are denoted $(\mathbf{e}_r, \mathbf{e}_\theta, \mathbf{e}_\phi)$. The unit vector normal to a surface is simply denoted $\mathbf{n}$.

1.3.1 Scalar and vector products

The scalar product of two vectors $\mathbf{a}$ and $\mathbf{b}$ is defined as the product of their magnitudes and the cosine of the angle α between the vectors:

$$\mathbf{a} \cdot \mathbf{b} = ab \cos \alpha \qquad (1.8)$$

If the vectors are orthogonal, the cosine of the angle α is zero and

$$\mathbf{a} \cdot \mathbf{b} = 0 \qquad (1.9)$$

The vector product of two vectors is another vector, whose direction is perpendicular to both vectors, such that a right-handed rule is observed. The magnitude of the vector product is the product of the individual vector magnitudes and the sine of the angle α between the vectors:

$$|\mathbf{a} \times \mathbf{b}| = ab \sin \alpha \qquad (1.10)$$

If $\mathbf{a}$ and $\mathbf{b}$ are parallel, the sine of the angle between them is zero and

$$\mathbf{a} \times \mathbf{b} = 0 \qquad (1.11)$$

Applying these rules to the unit vectors $(\mathbf{e}_x, \mathbf{e}_y, \mathbf{e}_z)$, which are normal to each other and have unit magnitude, it follows that their scalar products are

$$\begin{aligned} \mathbf{e}_x \cdot \mathbf{e}_y = \mathbf{e}_y \cdot \mathbf{e}_z = \mathbf{e}_z \cdot \mathbf{e}_x = 0 \\ \mathbf{e}_x \cdot \mathbf{e}_x = \mathbf{e}_y \cdot \mathbf{e}_y = \mathbf{e}_z \cdot \mathbf{e}_z = 1 \end{aligned} \qquad (1.12)$$

The vector products of the unit vectors are

$$\begin{aligned} \mathbf{e}_x \times \mathbf{e}_y &= \mathbf{e}_z \\ \mathbf{e}_y \times \mathbf{e}_z &= \mathbf{e}_x \\ \mathbf{e}_z \times \mathbf{e}_x &= \mathbf{e}_y \\ \mathbf{e}_x \times \mathbf{e}_x &= \mathbf{e}_y \times \mathbf{e}_y = \mathbf{e}_z \times \mathbf{e}_z = 0 \end{aligned} \qquad (1.13)$$

A vector **a** with components (a_x, a_y, a_z) is expressed in terms of the unit vectors $(\mathbf{e}_x, \mathbf{e}_y, \mathbf{e}_z)$ as

$$\mathbf{a} = a_x\mathbf{e}_x + a_y\mathbf{e}_y + a_z\mathbf{e}_z \tag{1.14}$$

The scalar product of the vectors **a** and **b** is found by applying the relationships in (1.12):

$$
\begin{aligned}
\mathbf{a} \cdot \mathbf{b} &= \left(a_x\mathbf{e}_x + a_y\mathbf{e}_y + a_z\mathbf{e}_z\right)\cdot\left(b_x\mathbf{e}_x + b_y\mathbf{e}_y + b_z\mathbf{e}_z\right) \\
&= a_xb_x + a_yb_y + a_zb_z
\end{aligned} \tag{1.15}
$$

The vector product of the vectors **a** and **b** is found by using (1.13):

$$
\begin{aligned}
\mathbf{a} \times \mathbf{b} &= \left(a_x\mathbf{e}_x + a_y\mathbf{e}_y + a_z\mathbf{e}_z\right) \times \left(b_x\mathbf{e}_x + b_y\mathbf{e}_y + b_z\mathbf{e}_z\right) \\
&= \left(a_yb_z - a_zb_y\right)\mathbf{e}_x + \left(a_zb_x - a_xb_z\right)\mathbf{e}_y + \left(a_xb_y - a_yb_x\right)\mathbf{e}_z
\end{aligned} \tag{1.16}
$$

This result leads to a convenient way of evaluating the vector product of two vectors, by writing their components as the elements of a determinant, as follows:

$$\mathbf{a} \times \mathbf{b} = \begin{vmatrix} \mathbf{e}_x & \mathbf{e}_y & \mathbf{e}_z \\ a_x & a_y & a_z \\ b_x & b_y & b_z \end{vmatrix} \tag{1.17}$$

The following relationships may be established, in a similar manner to the above, for combinations of scalar and vector products of the vectors **a**, **b**, and **c**:

$$\mathbf{a} \cdot (\mathbf{b} \times \mathbf{c}) = \mathbf{b} \cdot (\mathbf{c} \times \mathbf{a}) = \mathbf{c} \cdot (\mathbf{a} \times \mathbf{b}) \tag{1.18}$$

$$\mathbf{a} \times (\mathbf{b} \times \mathbf{c}) = \mathbf{b}(\mathbf{c} \cdot \mathbf{a}) - \mathbf{c}(\mathbf{a} \cdot \mathbf{b}) \tag{1.19}$$

$$(\mathbf{a} \times \mathbf{b}) \times \mathbf{c} = \mathbf{b}(\mathbf{c} \cdot \mathbf{a}) - \mathbf{a}(\mathbf{b} \cdot \mathbf{c}) \tag{1.20}$$

1.3.2 Vector differential operations

The vector differential operator ∇ is defined relative to Cartesian axes (x, y, z) as

$$\nabla = \mathbf{e}_x\frac{\partial}{\partial x} + \mathbf{e}_y\frac{\partial}{\partial y} + \mathbf{e}_z\frac{\partial}{\partial z} \tag{1.21}$$

The vector operator ∇ determines the gradient of a scalar function, which may be understood as the rate of change of the function in the direction of each of the reference axes. For example, the gradient of the scalar function φ with respect to Cartesian axes is the vector

$$\nabla \varphi = \mathbf{e}_x \frac{\partial \varphi}{\partial x} + \mathbf{e}_y \frac{\partial \varphi}{\partial y} + \mathbf{e}_z \frac{\partial \varphi}{\partial z} \tag{1.22}$$

The vector operator ∇ can operate on either a scalar quantity or a vector. The scalar product of ∇ with a vector is called the *divergence* of the vector. Applied to the vector $\mathbf{a}$ it is equal to

$$\nabla \cdot \mathbf{a} = \left(\mathbf{e}_x \frac{\partial}{\partial x} + \mathbf{e}_y \frac{\partial}{\partial y} + \mathbf{e}_z \frac{\partial}{\partial z} \right) \cdot (a_x \mathbf{e}_x + a_y \mathbf{e}_y + a_z \mathbf{e}_z)$$
$$= \frac{\partial a_x}{\partial x} + \frac{\partial a_y}{\partial y} + \frac{\partial a_z}{\partial z} \tag{1.23}$$

If the vector $\mathbf{a}$ is defined as the gradient of a scalar potential φ, as in (1.22), we can substitute potential gradients for the vector components (a_x, a_y, a_z). This gives

$$\nabla \cdot \nabla \varphi = \frac{\partial}{\partial x} \left(\frac{\partial \varphi}{\partial x} \right) + \frac{\partial}{\partial y} \left(\frac{\partial \varphi}{\partial y} \right) + \frac{\partial}{\partial z} \left(\frac{\partial \varphi}{\partial z} \right) \tag{1.24}$$

By convention the scalar product ($\nabla \cdot \nabla$) on the left is written ∇^2. The resulting identity is very important in potential theory and is encountered frequently. In Cartesian coordinates it is

$$\nabla^2 \varphi = \frac{\partial^2 \varphi}{\partial x^2} + \frac{\partial^2 \varphi}{\partial y^2} + \frac{\partial^2 \varphi}{\partial z^2} \tag{1.25}$$

The vector product of ∇ with a vector is called the *curl* of the vector. The curl of the vector $\mathbf{a}$ may be obtained using a determinant similar to (1.17):

$$\nabla \times \mathbf{a} = \begin{vmatrix} \mathbf{e}_x & \mathbf{e}_y & \mathbf{e}_z \\ \partial/\partial x & \partial/\partial y & \partial/\partial z \\ a_x & a_y & a_z \end{vmatrix} \tag{1.26}$$

In expanded format, this becomes

$$\nabla \times \mathbf{a} = \left(\frac{\partial a_z}{\partial y} - \frac{\partial a_y}{\partial z} \right) \mathbf{e}_x + \left(\frac{\partial a_x}{\partial z} - \frac{\partial a_z}{\partial x} \right) \mathbf{e}_y + \left(\frac{\partial a_y}{\partial x} - \frac{\partial a_x}{\partial y} \right) \mathbf{e}_z \tag{1.27}$$

The curl is sometimes called the *rotation* of a vector, because of its physical interpretation (Box 1.1). Some commonly encountered divergence and curl operations on combinations of the scalar quantity φ and the vectors $\mathbf{a}$ and $\mathbf{b}$ are listed below:

$$\nabla \cdot (\varphi \mathbf{a}) = (\nabla \varphi) \cdot \mathbf{a} + \varphi (\nabla \cdot \mathbf{a}) \tag{1.28}$$

Box 1.1.   **The curl of a vector**

The curl of a vector at a given point is related to the circulation of the vector about that point. This interpretation is best illustrated by an example, in which a fluid is rotating about a point with constant angular velocity $\boldsymbol{\omega}$. At distance $\mathbf{r}$ from the point the linear velocity of the fluid $\mathbf{v}$ is equal to $\boldsymbol{\omega} \times \mathbf{r}$. Taking the curl of $\mathbf{v}$, and applying the identity (1.31) with $\boldsymbol{\omega}$ constant,

$$\nabla \times \mathbf{v} = \nabla \times (\boldsymbol{\omega} \times \mathbf{r}) = \boldsymbol{\omega}(\nabla \cdot \mathbf{r}) - (\boldsymbol{\omega} \cdot \nabla)\mathbf{r} \qquad (1)$$

To evaluate the first term on the right, we use rectangular coordinates (x, y, z):

$$\boldsymbol{\omega}(\nabla \cdot \mathbf{r}) = \boldsymbol{\omega}\left(\mathbf{e}_x \frac{\partial}{\partial x} + \mathbf{e}_y \frac{\partial}{\partial y} + \mathbf{e}_z \frac{\partial}{\partial z}\right) \cdot (x\mathbf{e}_x + y\mathbf{e}_y + z\mathbf{e}_z)$$
$$= \boldsymbol{\omega}(\mathbf{e}_x \cdot \mathbf{e}_x + \mathbf{e}_y \cdot \mathbf{e}_y + \mathbf{e}_z \cdot \mathbf{e}_z) = 3\boldsymbol{\omega} \qquad (2)$$

The second term is

$$(\boldsymbol{\omega} \cdot \nabla)r = \left(\omega_x \frac{\partial}{\partial x} + \omega_y \frac{\partial}{\partial y} + \omega_z \frac{\partial}{\partial z}\right) \cdot (x\mathbf{e}_x + y\mathbf{e}_y + z\mathbf{e}_z)$$
$$= \omega_x \mathbf{e}_x + \omega_y \mathbf{e}_y + \omega_z \mathbf{e}_z = \boldsymbol{\omega} \qquad (3)$$

Combining the results gives

$$\nabla \times \mathbf{v} = 2\boldsymbol{\omega} \qquad (4)$$

$$\boldsymbol{\omega} = \frac{1}{2}(\nabla \times \mathbf{v}) \qquad (5)$$

Because of this relationship between the angular velocity and the linear velocity of a fluid, the curl operation is often interpreted as the *rotation* of the fluid. When $\nabla \times \mathbf{v} = 0$ everywhere, there is no rotation. A vector that satisfies this condition is said to be *irrotational*.

$$\nabla \cdot (\mathbf{a} \times \mathbf{b}) = \mathbf{b} \cdot (\nabla \times \mathbf{a}) - \mathbf{a} \cdot (\nabla \times \mathbf{b}) \qquad (1.29)$$

$$\nabla \times (\varphi \mathbf{a}) = (\nabla \varphi) \times \mathbf{a} + \varphi(\nabla \times \mathbf{a}) \qquad (1.30)$$

$$\nabla \times (\mathbf{a} \times \mathbf{b}) = \mathbf{a}(\nabla \cdot \mathbf{b}) - \mathbf{b}(\nabla \cdot \mathbf{a}) - (\mathbf{a} \cdot \nabla)\mathbf{b} + (\mathbf{b} \cdot \nabla)\mathbf{a} \qquad (1.31)$$

$$\nabla \times (\nabla \varphi) = 0 \qquad (1.32)$$

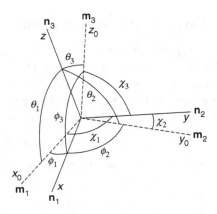

Fig. 1.3. Two sets of Cartesian coordinate axes, (x, y, z) and (x_0, y_0, z_0), with corresponding unit vectors $(\mathbf{n}_1, \mathbf{n}_2, \mathbf{n}_3)$ and $(\mathbf{m}_1, \mathbf{m}_2, \mathbf{m}_3)$, rotated relative to each other.

$$\nabla \cdot (\nabla \times \mathbf{a}) = 0 \tag{1.33}$$

$$\nabla \times (\nabla \times \mathbf{a}) = \nabla(\nabla \cdot \mathbf{a}) - \nabla^2 \mathbf{a} \tag{1.34}$$

It is a worthwhile exercise to establish these identities from basic principles, especially (1.19) and (1.31)–(1.34), which will be used in later chapters.

1.4 Matrices and tensors

1.4.1 The rotation matrix

Consider two sets of orthogonal Cartesian coordinate axes (x, y, z) and (x_0, y_0, z_0) that are inclined to each other as in Fig. 1.3. The x_0-axis makes angles $(\phi_1, \chi_1, \theta_1)$ with each of the (x, y, z) axes in turn. Similar sets of angles $(\phi_2, \chi_2, \theta_2)$ and $(\phi_3, \chi_3, \theta_3)$ are defined by the orientations of the y_0- and z_0-axes, respectively, to the (x, y, z) axes. Let the unit vectors along the (x, y, z) and (x_0, y_0, z_0) axes be $(\mathbf{n}_1, \mathbf{n}_2, \mathbf{n}_3)$ and $(\mathbf{m}_1, \mathbf{m}_2, \mathbf{m}_3)$, respectively. The vector $\mathbf{r}$ can be expressed in either system, i.e., $\mathbf{r} = \mathbf{r}(x, y, z) = \mathbf{r}(x_0, y_0, z_0)$, or, in terms of the unit vectors,

$$\mathbf{r} = x\mathbf{n}_1 + y\mathbf{n}_2 + z\mathbf{n}_3 = x_0\mathbf{m}_1 + y_0\mathbf{m}_2 + z_0\mathbf{m}_3 \tag{1.35}$$

We can write the scalar product $(\mathbf{r} \cdot \mathbf{m}_1)$ as

$$\mathbf{r} \cdot \mathbf{m}_1 = x\mathbf{n}_1 \cdot \mathbf{m}_1 + y\mathbf{n}_2 \cdot \mathbf{m}_1 + z\mathbf{n}_3 \cdot \mathbf{m}_1 = x_0 \tag{1.36}$$

The scalar product $(\mathbf{n}_1 \cdot \mathbf{m}_1) = \cos\phi_1 = \alpha_{11}$ defines α_{11} as the *direction cosine* of the x_0-axis with respect to the x-axis (Box 1.2). Similarly, $(\mathbf{n}_2 \cdot \mathbf{m}_1) = \cos\chi_1 = \alpha_{12}$

and $(\mathbf{n}_3 \cdot \mathbf{m}_1) = \cos\theta_1 = \alpha_{13}$ define α_{12} and α_{13} as the direction cosines of the x_0-axis with respect to the y- and z-axes, respectively. Thus, (1.36) is equivalent to

$$x_0 = \alpha_{11}x + \alpha_{12}y + \alpha_{13}z \tag{1.37}$$

On treating the y_0- and z_0-axes in the same way, we get their relationships to the (x, y, z) axes:

$$y_0 = \alpha_{21}x + \alpha_{22}y + \alpha_{23}z$$
$$z_0 = \alpha_{31}x + \alpha_{32}y + \alpha_{33}z \tag{1.38}$$

The three equations can be written as a single matrix equation

$$\begin{bmatrix} x_0 \\ y_0 \\ z_0 \end{bmatrix} = \begin{bmatrix} \alpha_{11} & \alpha_{12} & \alpha_{13} \\ \alpha_{21} & \alpha_{22} & \alpha_{23} \\ \alpha_{31} & \alpha_{32} & \alpha_{33} \end{bmatrix} \begin{bmatrix} x \\ y \\ z \end{bmatrix} = M \begin{bmatrix} x \\ y \\ z \end{bmatrix} \tag{1.39}$$

The coefficients α_{nm} ($n = 1, 2, 3$; $m = 1, 2, 3$) are the cosines of the interaxial angles. By definition, $\alpha_{12} = \alpha_{21}$, $\alpha_{23} = \alpha_{32}$, and $\alpha_{31} = \alpha_{13}$, so the square matrix M is symmetric. It transforms the components of the vector in the (x, y, z) coordinate system to corresponding values in the (x_0, y_0, z_0) coordinate system. It is thus equivalent to a rotation of the reference axes.

Because of the orthogonality of the reference axes, useful relationships exist between the direction cosines, as shown in Box 1.2. For example,

$$(\alpha_{11})^2 + (\alpha_{12})^2 + (\alpha_{13})^2 = \cos^2\phi_1 + \cos^2\chi_1 + \cos^2\theta_1 = \frac{1}{r^2}\left(x^2 + y^2 + z^2\right) = 1 \tag{1.40}$$

and

$$\alpha_{11}\alpha_{21} + \alpha_{12}\alpha_{22} + \alpha_{13}\alpha_{23} = \cos\phi_1\cos\phi_2 + \cos\chi_1\cos\chi_2 + \cos\theta_1\cos\theta_2 = 0 \tag{1.41}$$

The last summation is zero because it is the cosine of the right angle between the x_0-axis and the y_0-axis.

These two results can be summarized as

$$\sum_{k=1}^{3} \alpha_{mk}\alpha_{nk} = \begin{cases} 1, & m = n \\ 0, & m \neq n \end{cases} \tag{1.42}$$

1.4.2 Eigenvalues and eigenvectors

The transpose of a matrix X with elements α_{nm} is a matrix with elements α_{mn} (i.e., the elements in the rows are interchanged with corresponding elements in

Box 1.2. **Direction cosines**

The vector $\mathbf{r}$ is inclined at angles α, β, and γ, respectively, to orthogonal reference axes (x, y, z) with corresponding unit vectors $(\mathbf{e}_x, \mathbf{e}_y, \mathbf{e}_z)$, as in Fig. B1.2. The vector $\mathbf{r}$ can be written

$$\mathbf{r} = x\mathbf{e}_x + y\mathbf{e}_y + z\mathbf{e}_z \tag{1}$$

where (x, y, z) are the components of $\mathbf{r}$ with respect to these axes. The scalar products of $\mathbf{r}$ with $\mathbf{e}_x$, $\mathbf{e}_y$, and $\mathbf{e}_z$ are

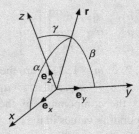

Fig. B1.2. Angles α, β, and γ define the tilt of a vector $\mathbf{r}$ relative to orthogonal reference axes (x, y, z), respectively. The unit vectors $(\mathbf{e}_x, \mathbf{e}_y, \mathbf{e}_z)$ define the coordinate system.

$$\mathbf{r} \cdot \mathbf{e}_x = x = r \cos \alpha$$
$$\mathbf{r} \cdot \mathbf{e}_y = y = r \cos \beta \tag{2}$$
$$\mathbf{r} \cdot \mathbf{e}_z = z = r \cos \gamma$$

Therefore, the vector $\mathbf{r}$ in (1) is equivalent to

$$\mathbf{r} = (r \cos \alpha)\mathbf{e}_x + (r \cos \beta)\mathbf{e}_y + (r \cos \gamma)\mathbf{e}_z \tag{3}$$

The unit vector $\mathbf{u}$ in the direction of $\mathbf{r}$ has the same direction as $\mathbf{r}$ but its magnitude is unity:

$$\mathbf{u} = \frac{\mathbf{r}}{r} = (\cos \alpha)\mathbf{e}_x + (\cos \beta)\mathbf{e}_y + (\cos \gamma)\mathbf{e}_z = l\mathbf{e}_x + m\mathbf{e}_y + n\mathbf{e}_z \tag{4}$$

where (l, m, n) are the cosines of the angles that the vector $\mathbf{r}$ makes with the reference axes, and are called the direction cosines of $\mathbf{r}$. They are useful for describing the orientations of lines and vectors.

The scalar product of two unit vectors is the cosine of the angle they form. Let $\mathbf{u}_1$ and $\mathbf{u}_2$ be unit vectors representing straight lines with direction cosines (l_1, m_1, n_1) and (l_2, m_2, n_2), respectively, and let θ be the angle between the vectors. The scalar product of the vectors is

$$\mathbf{u}_1 \cdot \mathbf{u}_2 = \cos\theta = \left(l_1\mathbf{e}_x + m_1\mathbf{e}_y + n_1\mathbf{e}_z\right) \cdot \left(l_2\mathbf{e}_x + m_2\mathbf{e}_y + n_2\mathbf{e}_z\right) \quad (5)$$

Therefore,

$$\cos\theta = l_1l_2 + m_1m_2 + n_1n_2 \quad (6)$$

The square of a unit vector is the scalar product of the vector with itself and is equal to 1:

$$\mathbf{u} \cdot \mathbf{u} = \frac{\mathbf{r} \cdot \mathbf{r}}{r^2} = 1 \quad (7)$$

On writing the unit vector $\mathbf{u}$ as in (4), and applying the orthogonality conditions from (2), we find that the sum of the squares of the direction cosines of a line is unity:

$$\left(l\mathbf{e}_x + m\mathbf{e}_y + n\mathbf{e}_z\right) \cdot \left(l\mathbf{e}_x + m\mathbf{e}_y + n\mathbf{e}_z\right) = l^2 + m^2 + n^2 = 1 \quad (8)$$

the columns). The transpose of a (3×1) column matrix is a (1×3) row matrix. For example, if X is a column matrix given by

$$X = \begin{bmatrix} x \\ y \\ z \end{bmatrix} \quad (1.43)$$

then its transpose is the row matrix X^T, where

$$X^\mathrm{T} = [x \quad y \quad z] \quad (1.44)$$

The matrix equation $X^\mathrm{T}MX = K$, where K is a constant, defines a *quadric* surface:

$$X^\mathrm{T}MX = [x \quad y \quad z] \begin{bmatrix} \alpha_{11} & \alpha_{12} & \alpha_{13} \\ \alpha_{21} & \alpha_{22} & \alpha_{23} \\ \alpha_{31} & \alpha_{32} & \alpha_{33} \end{bmatrix} \begin{bmatrix} x \\ y \\ z \end{bmatrix} = K \quad (1.45)$$

The symmetry of the matrix leads to the equation of this surface:

$$f(x, y, z) = a_{11}x^2 + a_{22}y^2 + a_{33}z^2 + 2a_{12}xy + 2a_{23}yz + 2a_{31}zx = K \quad (1.46)$$

When the coefficients a_{nm} are all positive real numbers, the geometric expression of the quadratic equation is an ellipsoid. The normal direction $\mathbf{n}$ to the surface of the ellipsoid at the point $P(x, y, z)$ is the gradient of the surface. Using the relationships between (x, y, z) and (x_0, y_0, z_0) in (1.39) and the symmetry of the rotation matrix, $a_{nm} = a_{mn}$ for $n \neq m$, the normal direction has components

$$\frac{\partial f}{\partial x} = 2(a_{11}x + a_{12}y + a_{13}z) = 2x_0$$

$$\frac{\partial f}{\partial y} = 2(a_{21}x + a_{22}y + a_{23}z) = 2y_0$$

$$\frac{\partial f}{\partial z} = 2(a_{31}x + a_{32}y + a_{33}z) = 2z_0 \quad (1.47)$$

and we write

$$\mathbf{n}(x, y, z) = \nabla f = \mathbf{e}_x \frac{\partial f}{\partial x} + \mathbf{e}_y \frac{\partial f}{\partial x} + \mathbf{e}_z \frac{\partial f}{\partial x} \quad (1.48)$$

$$\mathbf{n}(x, y, z) = 2(x_0\mathbf{e}_x + y_0\mathbf{e}_y + z_0\mathbf{e}_z) = 2\mathbf{r}(x_0, y_0, z_0) \quad (1.49)$$

The normal $\mathbf{n}$ to the surface at $P(x, y, z)$ in the original coordinates is parallel to the vector $\mathbf{r}$ at the point (x_0, y_0, z_0) in the rotated coordinates (Fig. 1.4).

The transformation matrix M has the effect of rotating the reference axes from one orientation to another. A particular matrix exists that will cause the directions of the (x_0, y_0, z_0) axes to coincide with the (x, y, z) axes. In this case the normal to the surface of the ellipsoid is one of the three principal axes of the

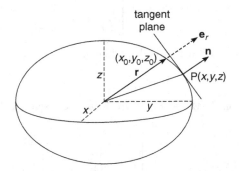

Fig. 1.4. Location of a point (x, y, z) on an ellipsoid, where the normal $\mathbf{n}$ to the surface is parallel to the radius vector at the point (x_0, y_0, z_0).

ellipsoid. The component x_0 is then proportional to x, y_0 is proportional to y, and z_0 is proportional to z. Let the proportionality constant be β. Then $x_0 = \beta x$, $y_0 = \beta y$, and $z_0 = \beta z$, and we get the set of simultaneous equations

$$
\begin{aligned}
(\alpha_{11} - \beta)x + \alpha_{12}y + \alpha_{13}z &= 0 \\
\alpha_{21}x + (\alpha_{22} - \beta)y + \alpha_{23}z &= 0 \\
\alpha_{31}x + \alpha_{32}y + (\alpha_{33} - \beta)z &= 0
\end{aligned}
\tag{1.50}
$$

which, in matrix form, is

$$
\begin{bmatrix}
\alpha_{11} - \beta & \alpha_{12} & \alpha_{13} \\
\alpha_{21} & \alpha_{22} - \beta & \alpha_{23} \\
\alpha_{31} & \alpha_{32} & \alpha_{33} - \beta
\end{bmatrix}
\begin{bmatrix}
x \\ y \\ z
\end{bmatrix} = 0
\tag{1.51}
$$

The simultaneous equations have a non-trivial solution only if the determinant of coefficients is zero, i.e.,

$$
\begin{vmatrix}
\alpha_{11} - \beta & \alpha_{12} & \alpha_{13} \\
\alpha_{21} & \alpha_{22} - \beta & \alpha_{23} \\
\alpha_{31} & \alpha_{32} & \alpha_{33} - \beta
\end{vmatrix} = 0
\tag{1.52}
$$

This equation is a third-order polynomial in β. Its three roots $(\beta_1, \beta_2, \beta_3)$ are known as the *eigenvalues* of the matrix M. When each eigenvalue β_n is inserted in turn into (1.50) it defines the components of a corresponding vector $\mathbf{v}_n$, which is called an *eigenvector* of M.

Note that (1.51) is equivalent to the matrix equation

$$
\begin{bmatrix}
\alpha_{11} & \alpha_{12} & \alpha_{13} \\
\alpha_{21} & \alpha_{22} & \alpha_{23} \\
\alpha_{31} & \alpha_{32} & \alpha_{33}
\end{bmatrix}
\begin{bmatrix}
x \\ y \\ z
\end{bmatrix}
- \beta
\begin{bmatrix}
1 & 0 & 0 \\
0 & 1 & 0 \\
0 & 0 & 1
\end{bmatrix}
\begin{bmatrix}
x \\ y \\ z
\end{bmatrix} = 0
\tag{1.53}
$$

which we can write in symbolic form

$$
(M - \beta I)X = 0
\tag{1.54}
$$

The matrix I, with diagonal elements equal to 1 and off-diagonal elements 0, is called a unit matrix:

$$
I =
\begin{bmatrix}
1 & 0 & 0 \\
0 & 1 & 0 \\
0 & 0 & 1
\end{bmatrix}
\tag{1.55}
$$

1.4.3 Tensor notation

Equations describing vector relationships can become cumbersome when written in full or symbolic form. Tensor notation provides a succinct alternative way of writing the equations. Instead of the alphabetic indices used in the previous section, tensor notation uses numerical indices that allow summations to be expressed in a compact form.

Let the Cartesian coordinates (x, y, z) be replaced by coordinates (x_1, x_2, x_3) and let the corresponding unit vectors be (e_1, e_2, e_3). The vector **a** in (1.14) becomes

$$\mathbf{a} = a_1\mathbf{e}_1 + a_2\mathbf{e}_2 + a_3\mathbf{e}_3 = \sum_{i=1,2,3} a_i\mathbf{e}_i \qquad (1.56)$$

A convention introduced by Einstein drops the summation sign and tacitly assumes that repetition of an index implies summation over all values of the index, in this case from 1 to 3. The vector **a** is then written explicitly

$$\mathbf{a} = a_i\mathbf{e}_i \qquad (1.57)$$

Alternatively, the unit vectors can be implied and the expression a_i is understood to represent the vector **a**. Using the summation convention, (1.15) for the scalar product of two vectors **a** and **b** is

$$\mathbf{a} \cdot \mathbf{b} = a_1b_1 + a_2b_2 + a_3b_3 = a_ib_i \qquad (1.58)$$

Suppose that two vectors **a** and **b** are related, so that each component of **a** is a linear combination of the components of **b**. The relationship can be expressed in tensor notation as

$$a_i = T_{ij}b_j \qquad (1.59)$$

The indices i and j identify components of the vectors **a** and **b**; each index takes each of the values 1, 2, and 3 in turn. The quantity T_{ij} is a second-order (or second-rank) *tensor*, representing the array of nine coefficients (i.e., 3^2). A vector has three components (i.e., 3^1) and is a first-order tensor; a scalar property has a single (i.e., 3^0) value, its magnitude, and is a zeroth-order tensor.

To write the cross product of two vectors we need to define a new quantity, the Levi-Civita *permutation tensor* ε_{ijk}. It has the value $+1$ when a permutation of the indices is *even* (i.e., $\varepsilon_{123} = \varepsilon_{231} = \varepsilon_{312} = 1$) and the value -1 when a permutation of the indices is *odd* (i.e., $\varepsilon_{132} = \varepsilon_{213} = \varepsilon_{321} = -1$). If any pair of indices is equal, $\varepsilon_{ijk} = 0$. This enables us to write the cross product of two vectors in tensor notation. Let **u** be the cross product of vectors **a** and **b**:

$$\mathbf{u} = \mathbf{a} \times \mathbf{b} = (a_2b_3 - a_3b_2)\mathbf{e}_1 + (a_3b_1 - a_1b_3)\mathbf{e}_2 + (a_1b_2 - a_2b_1)\mathbf{e}_3 \quad (1.60)$$

In tensor notation this is written

$$u_i = \varepsilon_{ijk} a_j b_k \tag{1.61}$$

This can be verified readily for each component of **u**. For example,

$$u_1 = \varepsilon_{123} a_2 b_3 + \varepsilon_{132} a_3 b_2 = a_2 b_3 - a_3 b_2 \tag{1.62}$$

The tensor equivalent to the unit matrix defined in (1.55) is known as *Kronecker's symbol*, δ_{ij}, or alternatively the *Kronecker delta*. It has the values

$$\delta_{ij} = \begin{cases} 1, & \text{if } i = j \\ 0, & \text{if } i \neq j \end{cases} \tag{1.63}$$

Kronecker's symbol is convenient for selecting a particular component of a tensor equation. For example, (1.54) can be written in tensor form using the Kronecker symbol:

$$\left(\alpha_{ij} - \beta \delta_{ij}\right) x_j = 0 \tag{1.64}$$

This represents the set of simultaneous equations in (1.50). Likewise, the relationship between direction cosines in (1.42) simplifies to

$$\alpha_{mk} \alpha_{nk} = \delta_{mn} \tag{1.65}$$

in which a summation over the repeated index is implied.

1.4.4 Rotation of coordinate axes

Let v_k be a vector related to the coordinates x_l by the tensor T_{kl}

$$v_k = T_{kl} x_l \tag{1.66}$$

A second set of coordinates x'_n is rotated relative to the axes x_l so that the direction cosines of the angles between corresponding axes are the elements of the tensor α_{nl}:

$$x'_n = \alpha_{nl} x_l \tag{1.67}$$

Let the same vector be related to the rotated coordinate axes x'_n by the tensor T'_{kn}:

$$v'_k = T'_{kn} x'_n \tag{1.68}$$

v_k and v'_k are the same vector, expressed relative to different sets of axes. Therefore,

$$v'_k = \alpha_{kn} v_n = \alpha_{kn} T_{nl} x_l \tag{1.69}$$

Equating the expressions in (1.68) and (1.69) for v'_k gives

$$T'_{kn}x'_n = \alpha_{kn}T_{nl}x_l \tag{1.70}$$

Using the relationships between the axes in (1.67),

$$T'_{kn}x'_n = T'_{kn}\alpha_{nl}x_l \tag{1.71}$$

Therefore,

$$T'_{kn}\alpha_{nl} = \alpha_{kn}T_{nl} \tag{1.72}$$

On multiplying by α_{ml} and summing,

$$\alpha_{ml}\alpha_{nl}T'_{kn} = \alpha_{ml}\alpha_{kn}T_{nl} \tag{1.73}$$

Note that in expanded form the products of direction cosines on the left are equal to

$$\alpha_{ml}\alpha_{nl} = \alpha_{m1}\alpha_{n1} + \alpha_{m2}\alpha_{n2} + \alpha_{m3}\alpha_{n3} = \delta_{mn} \tag{1.74}$$

as a result of (1.42). Therefore the transformation matrix in the rotated coordinate system is related to the original matrix by the direction cosines between the two sets of axes:

$$T'_{km} = \alpha_{ml}\alpha_{kn}T_{nl} \tag{1.75}$$

The indices m and k can be interchanged without affecting the result. The sequence of terms in the summation changes, but its sum does not. Therefore,

$$T'_{km} = \alpha_{kl}\alpha_{mn}T_{nl} \tag{1.76}$$

This relationship allows us to compute the elements of a matrix in a new coordinate system that is rotated relative to the original reference axes by angles that have the set of direction cosines α_{nl}.

1.4.5 Vector differential operations in tensor notation

In tensor notation the vector differential operator ∇ in Cartesian coordinates becomes

$$\nabla = \mathbf{e}_i \frac{\partial}{\partial x_i} \tag{1.77}$$

The *gradient* of a scalar function φ with respect to Cartesian unit vectors $(\mathbf{e}_1, \mathbf{e}_2, \mathbf{e}_3)$ is therefore

$$\nabla\varphi = \mathbf{e}_1 \frac{\partial\varphi}{\partial x_1} + \mathbf{e}_2 \frac{\partial\varphi}{\partial x_2} + \mathbf{e}_3 \frac{\partial\varphi}{\partial x_3} = \mathbf{e}_i \frac{\partial\varphi}{\partial x_i} \tag{1.78}$$

Several shorthand forms of this equation are in common use; for example,

$$\frac{\partial\varphi}{\partial x_i} = (\nabla\varphi)_i = \varphi_{,i} = \partial_i\varphi \tag{1.79}$$

The *divergence* of the vector $\mathbf{a}$ is written in tensor notation as

$$\nabla\cdot\mathbf{a} = \frac{\partial a_1}{\partial x_1} + \frac{\partial a_2}{\partial x_2} + \frac{\partial a_3}{\partial x_3} = \frac{\partial a_i}{\partial x_i} = \partial_i a_i \tag{1.80}$$

The *curl* (or *rotation*) of the vector $\mathbf{a}$ becomes

$$\nabla\times\mathbf{a} = \mathbf{e}_1\left(\frac{\partial a_3}{\partial x_2} - \frac{\partial a_2}{\partial x_3}\right) + \mathbf{e}_2\left(\frac{\partial a_1}{\partial x_3} - \frac{\partial a_3}{\partial x_1}\right) + \mathbf{e}_3\left(\frac{\partial a_2}{\partial x_1} - \frac{\partial a_1}{\partial x_2}\right) \tag{1.81}$$

$$(\nabla\times\mathbf{a})_i = \varepsilon_{ijk}\frac{\partial a_k}{\partial x_j} = \varepsilon_{ijk}\partial_j a_k \tag{1.82}$$

1.5 Conservative force, field, and potential

If the work done in moving an object from one point to another against a force is independent of the path between the points, the force is said to be conservative. No work is done if the end-point of the motion is the same as the starting point; this condition is called the closed-path test of whether a force is conservative. In a real situation, energy may be lost, for example to heat or friction, but in an ideal case the total energy E is constant. The work dW done against the force $\mathbf{F}$ is converted into a gain dE_P in the potential energy of the displaced object. The change in the total energy dE is zero:

$$dE = dE_P + dW = 0 \tag{1.83}$$

The change in potential energy when a force with components (F_x, F_y, F_z) parallel to the respective Cartesian coordinate axes (x, y, z) experiences elementary displacements (dx, dy, dz) is

$$dE_P = -dW = -\left(F_x\,dx + F_y\,dy + F_z\,dz\right) \tag{1.84}$$

The value of a physical force may vary in the space around its source. For example, gravitational and electrical forces decrease with distance from a source mass or electrical charge, respectively. The region in which a physical quantity exerts a force is called its *field*. Its geometry is defined by lines tangential to the force at any point in the region. The term *field* is also used to express the value of the force exerted on a *unit* of the quantity. For example, the *electric field* of a charge is the force experienced by a unit charge at a given point; the *gravitational field* of a mass is the force acting on a unit of mass; it is therefore equivalent to the *acceleration*.

In a gravitational field the force **F** is proportional to the acceleration **a**. The Cartesian components of **F** are therefore (ma_x, ma_y, ma_z). The *gravitational potential U* is defined as the potential energy of a unit mass in the gravitational field, thus $dE_P = m\, dU$. After substituting these expressions into (1.84) we get

$$dU = -(a_x\, dx + a_y\, dy + a_z\, dz) \qquad (1.85)$$

The total differential dU can be written in terms of partial differentials as

$$dU = \frac{\partial U}{\partial x} dx + \frac{\partial U}{\partial y} dy + \frac{\partial U}{\partial z} dz \qquad (1.86)$$

On equating coefficients of dx, dy, and dz in these equations:

$$a_x = -\frac{\partial U}{\partial x}; \qquad a_y = -\frac{\partial U}{\partial y}; \qquad a_z = -\frac{\partial U}{\partial z} \qquad (1.87)$$

These relationships show that the acceleration **a** is the negative gradient of a scalar potential U:

$$\mathbf{a} = -\nabla U \qquad (1.88)$$

Similarly, other conservative fields (e.g., electric, magnetostatic) can be derived as the gradient of the corresponding scalar potential. According to the vector identity (1.32) the curl of a gradient is always zero; it follows from (1.88) that a conservative force-field **F** satisfies the condition

$$\nabla \times \mathbf{F} = 0 \qquad (1.89)$$

1.6 The divergence theorem (Gauss's theorem)

Let **n** be the unit vector normal to a surface element of area dS. The flux $d\Phi$ of a vector **F** across the surface element dS (Fig. 1.5) is defined to be the scalar product

$$d\Phi = \mathbf{F} \cdot \mathbf{n}\, dS \qquad (1.90)$$

If the angle between **F** and **n** is θ, the flux across dS is

$$d\Phi = F\, dS \cos\theta \qquad (1.91)$$

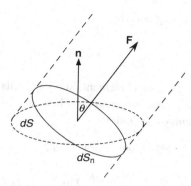

Fig. 1.5. The flux of a vector **F** across a small surface dS, whose normal **n** is inclined to the vector, is equal to the flux across a surface dS_n normal to the vector.

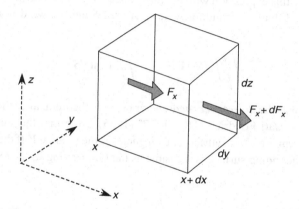

Fig. 1.6. Figure for computing the change in the flux of a vector in the x-direction for a small box with edges (dx, dy, dz).

where F is the magnitude of **F**. Thus the flux of **F** across the oblique surface dS is equivalent to that across the projection dS_n (=$dS \cos \theta$) of dS normal to **F**.

Consider the net flux of the vector **F** through a rectangular box with edges dx, dy, and dz parallel to the x-, y-, and z-axes, respectively (Fig. 1.6). The area dS_x of a side normal to the x-axis equals $dy\,dz$. The x-component of the vector at x, where it enters the box, is F_x, and at $x + dx$, where it leaves the box, it is $F_x + dF_x$. The net flux in the x-direction is

$$d\Phi_x = ((F_x + dF_x) - F_x)dS_x = dF_x\,dy\,dz \qquad (1.92)$$

If the distance dx is very small, the change in F_x may be written to first order as

$$dF_x = \frac{\partial F_x}{\partial x}dx \qquad (1.93)$$

The net flux in the x-direction is therefore

$$d\Phi_x = \frac{\partial F_x}{\partial x} dx\, dy\, dz = \frac{\partial F_x}{\partial x} dV \qquad (1.94)$$

where dV is the volume of the small element. Similar results are obtained for the net flux in each of the y- and z-directions. The total flux of **F** through the rectangular box is the sum of these flows:

$$d\Phi = d\Phi_x + d\Phi_y + d\Phi_z \qquad (1.95)$$

$$d\Phi = \left(\frac{\partial F_x}{\partial x} + \frac{\partial F_y}{\partial y} + \frac{\partial F_z}{\partial z}\right) dV = (\nabla \cdot \mathbf{F}) dV \qquad (1.96)$$

We can equate this expression with the flux defined in (1.90). The flux through a finite volume V with a bounding surface of area S and outward normal unit vector **n** is

$$\iiint_V (\nabla \cdot \mathbf{F}) dV = \iint_S \mathbf{F} \cdot \mathbf{n}\, dS \qquad (1.97)$$

This is known as the *divergence theorem*, or Gauss's theorem, after the German mathematician Carl Friedrich Gauss (1777–1855). Note that the surface S in Gauss's theorem is a *closed* surface, i.e., it encloses the volume V. If the flux of **F** entering the bounding surface is the same as the flux leaving it, the total flux is zero, and so

$$\nabla \cdot \mathbf{F} = 0 \qquad (1.98)$$

This is sometimes called the *continuity condition* because it implies that flux is neither created nor destroyed (i.e., there are neither sources nor sinks of the vector) within the volume. The vector is said to be *solenoidal*.

1.7 The curl theorem (Stokes' theorem)

Stokes' theorem relates the surface integral of the curl of a vector to the circulation of the vector around a closed path bounding the surface. Let the vector **F** pass through a surface S which is divided into a grid of small elements (Fig. 1.7). The area of a typical surface element is dS and the unit vector **n** normal to the element specifies its orientation.

First, we evaluate the work done by **F** around one of the small grid elements, ABCD (Fig. 1.8). Along each segment of the path we need to consider only the

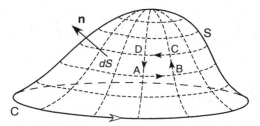

Fig. 1.7. Configuration for Stokes' theorem: the surface S is divided into a grid of elementary areas dS and is bounded by a closed circuit C.

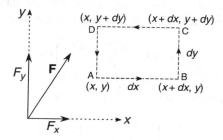

Fig. 1.8. Geometry for calculation of the work done by a force **F** around a small rectangular grid.

vector component parallel to that segment. The value of **F** may vary with position, so, for example, the x-component along AB may differ from the x-component along CD. Provided that dx and dy are infinitesimally small, we can use Taylor series approximations for the components of **F** (Section 1.10.2). To first order we get

$$
\begin{aligned}
(F_x)_{CD} &= (F_x)_{AB} + \frac{\partial F_x}{\partial y} dy \\
(F_y)_{BC} &= (F_y)_{DA} + \frac{\partial F_y}{\partial x} dx
\end{aligned}
\tag{1.99}
$$

The work done in a circuit around the small element ABCD is the sum of the work done along each individual segment:

$$
\oint_{ABCD} \mathbf{F} \cdot d\mathbf{l} = \int_x^{x+dx} (F_x)_{AB} \, dx + \int_y^{y+dy} (F_y)_{BC} \, dy + \int_{x+dx}^{x} (F_x)_{CD} \, dx + \int_{y+dy}^{y} (F_y)_{DA} \, dy
\tag{1.100}
$$

$$\oint_{\text{ABCD}} \mathbf{F} \cdot d\mathbf{l} = \int_{x}^{x+dx} \left((F_x)_{\text{AB}} - (F_x)_{\text{CD}} \right) dx + \int_{y}^{y+dy} \left((F_y)_{\text{BC}} - (F_y)_{\text{DA}} \right) dy$$

$$(1.101)$$

Substituting from (1.99) gives

$$\oint_{\text{ABCD}} \mathbf{F} \cdot d\mathbf{l} = \int_{x}^{x+dx} \left(-\frac{\partial F_x}{\partial y} dy \right) dx + \int_{y}^{y+dy} \left(\frac{\partial F_y}{\partial x} dx \right) dy \qquad (1.102)$$

The mean-value theorem allows us to replace the integrands over the tiny distances dx and dy by their values at some point in the range of integration:

$$\oint_{\text{ABCD}} \mathbf{F} \cdot d\mathbf{l} = \left(\frac{\partial F_y}{\partial x} - \frac{\partial F_x}{\partial y} \right) dx \, dy \qquad (1.103)$$

The bracketed expression is the z-component of the curl of $\mathbf{F}$

$$\oint_{\text{ABCD}} \mathbf{F} \cdot d\mathbf{l} = (\nabla \times \mathbf{F})_z \, dx \, dy \qquad (1.104)$$

The normal direction $\mathbf{n}$ to the small area $dS = dx \, dy$ is parallel to the z-axis (i.e., out of the plane of Fig. 1.8), and hence is in the direction of $(\nabla \times \mathbf{F})_z$. Thus,

$$\oint_{\text{ABCD}} \mathbf{F} \cdot d\mathbf{l} = (\nabla \times \mathbf{F}) \cdot n \, dS \qquad (1.105)$$

The circuit ABCD is one of many similar grid elements of the surface S. When adjacent elements are compared, the line integrals along their common boundary are equal and opposite. If the integration is carried out for the entire surface S, the only surviving parts are the integrations along the bounding curve C (Fig. 1.7). Thus

$$\iint_{S} (\nabla \times \mathbf{F}) \cdot n \, dS = \oint_{C} \mathbf{F} \cdot d\mathbf{l} \qquad (1.106)$$

This equation is known as Stokes' theorem, after the English mathematician George Gabriel Stokes (1819–1903). It enables conversion of the surface integral of a vector to a line integral. The integration on the left is made over the surface S through which the vector $\mathbf{F}$ passes. The closed integration on the right is made around the bounding curve C to the surface S; $d\mathbf{l}$ is an infinitesimal element of this boundary. The direction of $d\mathbf{l}$ around the curve is right-handed with respect to the surface S, i.e., positive when the path is kept to the right of the surface, as in Fig. 1.7.

Note that the surface S in Stokes' theorem is an *open* surface; it is like the surface of a bowl with the bounding curve C as its rim. The integration of **F** around the rim is called the circulation of **F** about the curve C. If the integral is zero, there is no circulation and the vector **F** is said to be *irrotational*. Comparison with the left-hand side shows that the condition for this is

$$\nabla \times \mathbf{F} = 0 \tag{1.107}$$

As shown in Section 1.5, this is also the condition for **F** to be a conservative field.

1.8 Poisson's equation

The derivations in this and the following sections are applicable to any field that varies as the inverse square of distance from its source. Gravitational acceleration is used as an example, but the electric field of a charge may be treated in the same way.

Let S be a surface enclosing an observer at P and a point mass m. Let dS be a small element of the surface at distance r in the direction $\mathbf{e}_r$ from the mass m, as in Fig. 1.9. The orientation of dS is specified by the direction **n** normal to the surface element. With G representing the gravitational constant (see Section 2.1), the gravitational acceleration $\mathbf{a}_G$ at dS is given by

$$\mathbf{a}_G = -G\frac{m}{r^2}\mathbf{e}_r \tag{1.108}$$

Let θ be the angle between the radius and the direction **n** normal to the surface element, and let the projection of dS normal to the radius be dS_n. The solid angle $d\Omega$ with apex at the mass is defined as the ratio of the normal surface element dS_n to the square of its distance r from the mass (Box 1.3):

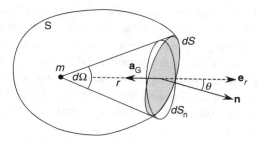

Fig. 1.9. Representation of the flux of the gravitational acceleration $\mathbf{a}_G$ through a closed surface S surrounding the source of the flux (the point mass m).

Box 1.3.   **Definition of a solid angle**

A small element of the surface of a sphere subtends a cone with apex at the center of the sphere (Fig. B1.3(a)). The solid angle Ω is defined as the ratio of the area A of the surface element to the square of the radius r of the sphere:

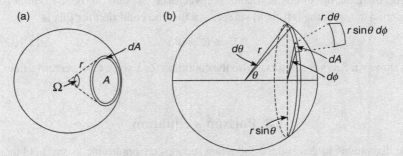

Fig. B1.3. (a) Relationship of the solid angle Ω, the area A of an element subtended on the surface of a sphere, and the radius r of the sphere. (b) The surface of a sphere divided into rings, and each ring into small surface elements with sides $r\,d\theta$ and $r\sin\theta\,d\phi$.

$$\Omega = \frac{A}{r^2} \tag{1}$$

This definition can be used for an arbitrarily shaped surface. If the surface is inclined to the radial direction it must be projected onto a surface normal to the radius, as in Fig. 1.5. For example, if the normal to the surface A makes an angle α with the direction from the apex of the subtended cone, the projected area is $A\cos\alpha$ and the solid angle Ω subtended by the area is

$$\Omega = \frac{A\cos\alpha}{r^2} \tag{2}$$

As an example, let the area on the surface of a sphere be enclosed by a small circle (Fig. B1.3(b)). Symmetry requires spherical polar coordinates to describe the area within the circle. Let the circle be divided into concentric rings, and let the half-angle subtended by a ring at the center of the sphere be θ. The radius of the ring is $r\sin\theta$ and its width is $r\,d\theta$. Let the angular position of a small surface element of the ring be ϕ; the length of a side of the element is then $r\sin\theta\,d\phi$. The area dA of a surface area element is equal to $r^2\sin\theta\,d\theta\,d\phi$. The solid angle subtended at the center of the sphere by the element of area dA is

$$d\Omega = \frac{r^2\sin\theta\,d\theta\,d\phi}{r^2} = \sin\theta\,d\theta\,d\phi \tag{3}$$

This expression is also equivalent to the element of area on the surface of a unit sphere (one with unit radius). Integrating in the ranges $0 \leq \phi \leq 2\pi$ and $0 \leq \theta \leq \theta_0$, we get the solid angle Ω_0 subtended by a circular region of the surface of a sphere defined by a half-apex angle θ_0:

$$\Omega_0 = \int\limits_{\theta=0}^{\theta_0} \int\limits_{\phi=0}^{2\pi} \sin\theta \, d\theta \, d\phi = 2\pi(1 - \cos\theta_0) \tag{4}$$

The unit of measurement of solid angle is the *steradian*, which is analogous to the radian in plane geometry. The maximum value of a solid angle is when the surface area is that of the complete sphere, namely $4\pi r^2$. The solid angle at its center then has the maximum possible value of 4π. This result is also obtained by letting the half-apex angle θ_0 in (4) increase to its maximum value π.

$$d\Omega = \frac{dS_n}{r^2} = \frac{dS \cos\theta}{r^2} = \frac{(\mathbf{e}_r \cdot \mathbf{n})dS}{r^2} \tag{1.109}$$

The flux dN of the gravitational acceleration $\mathbf{a}_G$ through the area element is

$$dN = \mathbf{a}_G \cdot \mathbf{n} \, dS = -G\frac{m}{r^2}(\mathbf{e}_r \cdot \mathbf{n})dS \tag{1.110}$$

$$dN = -Gm\frac{\cos\theta \, dS}{r^2} = -Gm \, d\Omega \tag{1.111}$$

If we integrate this expression over the entire surface S we get the total gravitational flux N,

$$N = \iint\limits_{S} \mathbf{a}_G \cdot \mathbf{n} \, dS = -\int\limits_{\Omega} Gm \, d\Omega = -4\pi Gm \tag{1.112}$$

Now we replace this surface integral by a volume integration, using the divergence theorem (Section 1.6)

$$\iiint\limits_{V} (\nabla \cdot \mathbf{a}_G)dV = \iint\limits_{S} \mathbf{a}_G \cdot \mathbf{n} \, dS = -4\pi Gm \tag{1.113}$$

This is valid for any point mass m inside the surface S. If the surface encloses many point masses we may replace m with the sum of the point masses. If mass is distributed in the volume with mean density ρ, a volume integral can replace the enclosed mass:

$$\iiint_V (\nabla \cdot \mathbf{a}_G)dV = -4\pi G \iiint_V \rho \, dV \tag{1.114}$$

$$\iiint_V (\nabla \cdot \mathbf{a}_G + 4\pi G\rho)dV = 0 \tag{1.115}$$

For this to be generally true the integrand must be zero. Consequently,

$$\nabla \cdot \mathbf{a}_G = -4\pi G\rho \tag{1.116}$$

The gravitational acceleration is the gradient of the gravitational potential U_G as in (1.88):

$$\nabla \cdot (-\nabla U_G) = -4\pi G\rho \tag{1.117}$$

$$\nabla^2 U_G = 4\pi G\rho \tag{1.118}$$

Equation (1.118) is known as Poisson's equation, after Siméon-Denis Poisson (1781–1840), a French mathematician and physicist. It describes the gravitational potential of a mass distribution at a point that is within the mass distribution. For example, it may be used to compute the gravitational potential at a point inside the Earth.

1.9 Laplace's equation

Another interesting case is the potential at a point outside the mass distribution. Let S be a closed surface outside the point mass m. The radius vector $\mathbf{r}$ from the point mass m now intersects the surface S at two points A and B, where it forms angles θ_1 and θ_2 with the respective unit vectors $\mathbf{n}_1$ and $\mathbf{n}_2$ normal to the surface (Fig. 1.10). Let $\mathbf{e}_r$ be a unit vector in the radial direction. Note that the outward normal $\mathbf{n}_1$ forms an obtuse angle with the radius vector at A. The gravitational acceleration at A is $\mathbf{a}_1$ and its flux through the surface area dS_1 is

$$dN_1 = \mathbf{a}_1 \cdot \mathbf{n}_1 \, dS_1 = \left(-\frac{Gm}{r_1^2}\right)(\mathbf{r} \cdot \mathbf{n}_1)dS_1 \tag{1.119}$$

$$dN_1 = \left(-\frac{Gm}{r_1^2}\right)\cos(\pi - \theta_1)dS_1 = Gm\frac{\cos\theta_1 \, dS_1}{r_1^2} \tag{1.120}$$

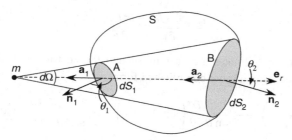

Fig. 1.10. Representation of the gravitational flux through a closed surface S that does not enclose the source of the flux (the point mass *m*).

$$dN_1 = Gm \, d\Omega \tag{1.121}$$

The gravitational acceleration at B is $\mathbf{a}_2$ and its flux through the surface area dS_2 is

$$dN_2 = \mathbf{a}_2 \cdot \mathbf{n}_2 \, dS_2 = -Gm\frac{\cos\theta_2 \, dS_2}{r_2^2} \tag{1.122}$$

$$dN_2 = -Gm \, d\Omega \tag{1.123}$$

The total contribution of both surfaces to the gravitational flux is

$$dN = dN_1 + dN_2 = 0 \tag{1.124}$$

Thus, the total flux of the gravitational acceleration $\mathbf{a}_G$ through a surface S that does not include the point mass *m* is zero. By invoking the divergence theorem we have for this situation

$$\int_V (\nabla \cdot \mathbf{a}_G) dV = \int_S \mathbf{a}_G \cdot \mathbf{n} \, dS = 0 \tag{1.125}$$

For this result to be valid for any volume, the integrand must be zero:

$$\nabla \cdot \mathbf{a}_G = \nabla \cdot (-\nabla U_G) = 0 \tag{1.126}$$

$$\nabla^2 U_G = 0 \tag{1.127}$$

Equation (1.127) is Laplace's equation, named after Pierre Simon, Marquis de Laplace (1749–1827), a French mathematician and physicist. It describes the gravitational potential at a point outside a mass distribution. For example, it is applicable to the computation of the gravitational potential of the Earth at an external point or on its surface.

In Cartesian coordinates, which are rectilinear, Laplace's equation has the simple form

$$\frac{\partial^2 U_G}{\partial x^2} + \frac{\partial^2 U_G}{\partial y^2} + \frac{\partial^2 U_G}{\partial z^2} = 0 \qquad (1.128)$$

Spherical polar coordinates are curvilinear and the curvature of the angular coordinates results in a more complicated form:

$$\frac{1}{r^2}\frac{\partial}{\partial r}\left(r^2\frac{\partial U_G}{\partial r}\right) + \frac{1}{r^2\sin\theta}\frac{\partial}{\partial\theta}\left(\sin\theta\,\frac{\partial U_G}{\partial\theta}\right) + \frac{1}{r^2\sin^2\theta}\frac{\partial^2 U_G}{\partial\phi^2} = 0 \quad (1.129)$$

1.10 Power series

A function $f(x)$ that is continuous and has continuous derivatives may be approximated by the sum of an infinite series of powers of x. Many mathematical functions – e.g., sin x, cos x, exp(x), ln($1+x$) – fulfill these conditions of continuity and can be expressed as power series. This often facilitates the calculation of a value of the function. Three types of power series will be considered here: the MacLaurin, Taylor, and binomial series.

1.10.1 MacLaurin series

Let the function $f(x)$ be written as an infinite sum of powers of x:

$$f(x) = a_0 + a_1 x + a_2 x^2 + a_3 x^3 + a_4 x^4 + \cdots + a_n x^n + \cdots \qquad (1.130)$$

The coefficients a_n in this sum are constants. Differentiating (1.130) repeatedly with respect to x gives

$$\frac{df}{dx} = a_1 + 2a_2 x + 3a_3 x^2 + 4a_4 x^3 + \cdots + na_n x^{n-1} + \cdots$$

$$\frac{d^2f}{dx^2} = 2a_2 + (3\cdot 2)a_3 x + (4\cdot 3)a_4 x^2 + \cdots + n(n-1)a_n x^{n-2} + \cdots$$

$$\frac{d^3f}{dx^3} = (3\cdot 2)a_3 + (4\cdot 3\cdot 2)a_4 x + \cdots + n(n-1)(n-2)a_n x^{n-3} + \cdots$$

$$\qquad (1.131)$$

After n differentiations, the expression becomes

$$\frac{d^nf}{dx^n} = (n(n-1)(n-2)\ldots 3\cdot 2\cdot 1)\,a_n + \text{terms containing powers of } x$$

$$\qquad (1.132)$$

Now we evaluate each of the differentiations at $x = 0$. Terms containing powers of x are zero and

$$f(0) = a_0, \qquad \left(\frac{df}{dx}\right)_{x=0} = a_1$$

$$\left(\frac{d^2f}{dx^2}\right)_{x=0} = 2a_2, \qquad \left(\frac{d^3f}{dx^3}\right)_{x=0} = (3\cdot2)a_3 \qquad (1.133)$$

$$\left(\frac{d^nf}{dx^n}\right)_{x=0} = (n(n-1)(n-2)\ldots3\cdot2\cdot1)a_n = n!a_n$$

On inserting these values for the coefficients into (1.130) we get the power series for $f(x)$:

$$f(x) = f(0) + x\left(\frac{df}{dx}\right)_{x=0} + \frac{x^2}{2!}\left(\frac{d^2f}{dx^2}\right)_{x=0} + \frac{x^3}{3!}\left(\frac{d^3f}{dx^3}\right)_{x=0} + \cdots$$
$$+ \frac{x^n}{n!}\left(\frac{d^nf}{dx^n}\right)_{x=0} + \cdots \qquad (1.134)$$

This is the MacLaurin series for $f(x)$ about the origin, $x = 0$. It was derived in the eighteenth century by the Scottish mathematician Colin MacLaurin (1698–1746) as a special case of a Taylor series.

The MacLaurin series is a convenient way to derive series expressions for several important functions. In particular,

$$\sin x = x - \frac{x^3}{3!} + \frac{x^5}{5!} - \frac{x^7}{7!} + \cdots (-1)^{n-1}\frac{x^{2n-1}}{(2n-1)!}\cdots$$

$$\cos x = 1 - \frac{x^2}{2!} + \frac{x^4}{4!} - \frac{x^6}{6!} + \cdots (-1)^{n-1}\frac{x^{2n-2}}{(2n-2)!}\cdots$$

$$\exp(x) = e^x = 1 + x + \frac{x^2}{2!} + \frac{x^3}{3!} + \cdots + \frac{x^{n-1}}{(n-1)!} + \cdots \qquad (1.135)$$

$$\ln(1+x) = \log_e(1+x) = x - \frac{x^2}{2} + \frac{x^3}{3} - \frac{x^4}{4} + \cdots (-1)^{n-1}\frac{x^n}{n}\cdots$$

1.10.2 Taylor series

We can write the power series in (1.134) for $f(x)$ centered on any new origin, for example $x = x_0$. To do this we substitute $(x - x_0)$ for x in the above derivation. The power series becomes

$$f(x) = f(x_0) + (x - x_0)\left(\frac{df}{dx}\right)_{x=x_0} + \frac{(x-x_0)^2}{2!}\left(\frac{d^2f}{dx^2}\right)_{x=x_0}$$
$$+ \frac{(x-x_0)^3}{3!}\left(\frac{d^3f}{dx^3}\right)_{x=x_0} + \cdots + \frac{(x-x_0)^n}{n!}\left(\frac{d^nf}{dx^n}\right)_{x=x_0} + \cdots$$
$$(1.136)$$

This is called a Taylor series, after an English mathematician, Brooks Taylor (1685–1731), who described its properties in 1712.

The MacLaurin and Taylor series are both approximations to the function $f(x)$. The remainder between the true function and its power series is a measure of how well the function is expressed by the series.

1.10.3 Binomial series

Finite series

An important series is the expansion of the function $f(x) = (a + x)^n$. If n is a positive integer, the expansion of $f(x)$ is a finite series, terminating after $(n+1)$ terms. Evaluating the series for some low values of n gives the following:

$$
\begin{aligned}
&n = 0: (a+x)^0 = 1 \\
&n = 1: (a+x)^1 = a+x \\
&n = 2: (a+x)^2 = a^2 + 2ax + x^2 \\
&n = 3: (a+x)^3 = a^3 + 3a^2x + 3ax^2 + x^3 \\
&n = 4: (a+x)^4 = a^4 + 4a^3x + 6a^2x^2 + 4ax^3 + x^4
\end{aligned}
\tag{1.137}
$$

The general expansion of $f(x)$ is therefore

$$
\begin{aligned}
(a+x)^n = a^n &+ na^{n-1}x + \frac{n(n-1)}{1\cdot 2}a^{n-2}x^2 + \cdots \\
&+ \frac{n(n-1)\ldots(n-k+1)}{k!}a^{n-k}x^k \cdots + x^n
\end{aligned}
\tag{1.138}
$$

The coefficient of the general kth term is equivalent to

$$
\frac{n(n-1)\ldots(n-k+1)}{k!} = \frac{n!}{k!(n-k)!}
\tag{1.139}
$$

This is called the *binomial coefficient*.

When the constant a is equal to 1 and n is a positive integer, we have the useful series expansion

$$
(1+x)^n = \sum_{k=0}^{n} \frac{n!}{k!(n-k)!}x^k
\tag{1.140}
$$

Infinite series

If the exponent in (1.140) is not a positive integer, the series does not terminate, but is an infinite series. The series for $f(x) = (1+x)^p$, in which the exponent p is not a positive integer, may be derived as a MacLaurin series:

$$\left(\frac{df}{dx}\right)_{x=0} = \left(p(1+x)^{p-1}\right)_{x=0} = p$$

$$\left(\frac{d^2f}{dx^2}\right)_{x=0} = \left(p(p-1)(1+x)^{p-2}\right)_{x=0} = p(p-1)$$

$$\left(\frac{d^nf}{dx^n}\right)_{x=0} = (p(p-1)\ldots(p-n+1)(1+x)^{p-n})_{x=0} = p(p-1)\ldots(p-n+1)$$

$$(1.141)$$

On inserting these terms into (1.134), and noting that $f(0) = 1$, we get for the binomial series

$$(1+x)^p = 1 + px + \frac{p(p-1)}{1\cdot 2}x^2 + \frac{p(p-1)(p-2)}{1\cdot 2\cdot 3}x^3 + \cdots$$
$$+ \frac{p(p-1)\ldots(p-n+1)}{n!}x^n + \cdots$$

$$(1.142)$$

If the exponent p is not an integer, or p is negative, the series is convergent in the range $-1 < x < 1$.

1.10.4 Linear approximations

The variations in some physical properties over the surface of the Earth are small in relation to the main property. For example, the difference between the polar radius c and the equatorial radius a expressed as a fraction of the equatorial radius defines the flattening f, which is equal to 1/298. This results from deformation of the Earth by the centrifugal force of its own rotation, which, expressed as a fraction m of the gravitational force, is equal to 1/289. Both f and m are less than three thousandths of the main property, so f^2, m^2, and the product fm are of the order of nine parts in a million and, along with higher-order combinations, are negligible. Curtailing the expansion of small quantities at first order helps keep equations manageable without significant loss of geophysical information.

In the following chapters much use will be made of such linear approximation. It simplifies the form of mathematical functions and the usable part of the series described above. For example, for small values of x or $(x - x_0)$, the following first-order approximations may be used:

$$\sin x \approx x, \qquad \cos x \approx 1$$
$$\exp(x) \approx 1 + x, \qquad \ln(1+x) \approx x$$
$$(1+x)^p \approx 1 + px$$

$$(1.143)$$

$$f(x) \approx f(x_0) + (x - x_0)\left(\frac{df}{dx}\right)_{x=x_0}$$

1.11 Leibniz's rule

Assume that $u(x)$ and $v(x)$ are differentiable functions of x. The derivative of their product is

$$\frac{d}{dx}(u(x)v(x)) = u(x)\frac{dv(x)}{dx} + v(x)\frac{du(x)}{dx} \qquad (1.144)$$

If we define the operator $D = d/dx$, we obtain a shorthand form of this equation:

$$D(uv) = u\,Dv + v\,Du \qquad (1.145)$$

We can differentiate the product (uv) a second time by parts,

$$\begin{aligned} D^2(uv) &= D(D(uv)) = u\,D^2v + (Du)(Dv) + (Dv)(Du) + v\,D^2u \\ &= u\,D^2v + 2(Du)(Dv) + v\,D^2u \end{aligned} \qquad (1.146)$$

and, continuing in this way,

$$\begin{aligned} D^3(uv) &= u\,D^3v + 3(Du)(D^2v) + 3(D^2u)(Dv) + v\,D^3u \\ D^4(uv) &= u\,D^4v + 4(Du)(D^3v) + 6(D^2u)(D^2v) + 4(D^2u)(Dv) + v\,D^4u \end{aligned} \qquad (1.147)$$

The coefficients in these equations are the binomial coefficients, as defined in (1.139). Thus after n differentiations we have

$$D^n(uv) = \sum_{k=0}^{n} \frac{n!}{k!(n-k)!}\left(D^k u\right)\left(D^{n-k} v\right) \qquad (1.148)$$

This relationship is known as Leibniz's rule, after Gottfried Wilhelm Leibniz (1646–1716), who invented infinitesimal calculus contemporaneously with Isaac Newton (1642–1727); each evidently did so independently of the other.

1.12 Legendre polynomials

Let r and R be the sides of a triangle that enclose an angle θ and let u be the side opposite this angle (Fig. 1.11). The angle and sides are related by the cosine rule

$$u^2 = r^2 + R^2 - 2rR\cos\theta \qquad (1.149)$$

Inverting this expression and taking the square root gives

$$\frac{1}{u} = \frac{1}{R}\left[1 - 2\left(\frac{r}{R}\right)\cos\theta + \left(\frac{r}{R}\right)^2\right]^{-1/2} \qquad (1.150)$$

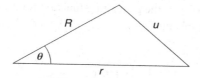

Fig. 1.11. Relationship of the sides r and R, which enclose an angle θ, and the side u opposite the angle, as used in the definition of Legendre polynomials.

Now let $h = r/R$ and $x = \cos\theta$, giving

$$\frac{1}{u} = \frac{1}{R}\left(1 - 2xh + h^2\right)^{-1/2} = \frac{1}{R}(1 - t)^{-1/2} \tag{1.151}$$

where $t = 2xh - h^2$. The equation can be expanded as a binomial series

$$(1-t)^{-1/2} = 1 + \left(-\frac{1}{2}\right)(-t) + \frac{\left(-\frac{1}{2}\right)\left(-\frac{3}{2}\right)}{1\cdot 2}(-t)^2 + \frac{\left(-\frac{1}{2}\right)\left(-\frac{3}{2}\right)\left(-\frac{5}{2}\right)}{1\cdot 2\cdot 3}(-t)^3 + \cdots$$

$$= 1 + \frac{t}{2} + \frac{1\cdot 3}{1\cdot 2}\left(\frac{t}{2}\right)^2 + \frac{1\cdot 3\cdot 5}{1\cdot 2\cdot 3}\left(\frac{t}{2}\right)^3 + \cdots$$

$$+ \frac{1\cdot 3\cdot 5\ldots(2n-1)}{1\cdot 2\cdot 3\ldots n}\left(\frac{t}{2}\right)^n + \cdots \tag{1.152}$$

The infinite series of terms on the right-hand side of the equation can be written

$$(1-t)^{-1/2} = \sum_{n=0}^{\infty} a_n t^n \tag{1.153}$$

The coefficient a_n is given by

$$a_n = \frac{1\cdot 3\cdot 5\cdot\ldots\cdot(2n-1)}{2^n n!} \tag{1.154}$$

Now, substitute the original expression for t,

$$\left(1 - 2xh + h^2\right)^{-1/2} = \sum_{n=0}^{\infty} a_n\left(2xh - h^2\right)^n = \sum_{n=0}^{\infty} a_n h^n(2x - h)^n \tag{1.155}$$

This equation is an infinite series in powers of h. The coefficient of each term in the power series is a polynomial in x. Let the coefficient of h^n be $P_n(x)$. The equation becomes

$$\Psi(x, h) = \left(1 - 2xh + h^2\right)^{-1/2} = \sum_{n=0}^{\infty} h^n P_n(x) \tag{1.156}$$

Equation (1.156) is known as the generating function for the polynomials $P_n(x)$. Using this result, and substituting $h = r/R$ and $x = \cos\theta$, we find that (1.151) becomes

$$\frac{1}{u} = \frac{1}{R}\sum_{n=0}^{\infty}\left(\frac{r}{R}\right)^n P_n(\cos\theta) \tag{1.157}$$

The polynomials $P_n(x)$ or $P_n(\cos\theta)$ are called Legendre polynomials, after the French mathematician Adrien-Marie Legendre (1752–1833). The defining equation (1.157) is called the *reciprocal-distance formula*. An alternative formulation is given in Box 1.4.

1.13 The Legendre differential equation

The Legendre polynomials satisfy an important second-order partial differential equation, which is called the Legendre differential equation. To derive this equation we will carry out a sequence of differentiations, starting with the generating function in the form

$$\Psi = \left(1 - 2xh + h^2\right)^{-1/2} \tag{1.158}$$

Differentiating this function once with respect to h gives

$$\frac{\partial\Psi}{\partial h} = (x - h)\left(1 - 2xh + h^2\right)^{-3/2} = (x - h)\Psi^3 \tag{1.159}$$

Differentiating Ψ twice with respect to x gives

$$\frac{\partial\Psi}{\partial x} = h\left(1 - 2xh + h^2\right)^{-3/2} = h\Psi^3$$
$$\Psi^3 = \frac{1}{h}\frac{\partial\Psi}{\partial x} \tag{1.160}$$

$$\frac{\partial^2\Psi}{\partial x^2} = 3h\Psi^2\frac{\partial\Psi}{\partial x} = 3h^2\Psi^5$$
$$\Psi^5 = \frac{1}{3h^2}\frac{\partial^2\Psi}{\partial x^2} \tag{1.161}$$

Next we perform successive differentiations of the product $(h\Psi)$ with respect to h. The first gives

$$\frac{\partial}{\partial h}(h\Psi) = \Psi + h\frac{\partial\Psi}{\partial h} = \Psi + h(x - h)\Psi^3 \tag{1.162}$$

Box 1.4. Alternative form of the reciprocal-distance formula

The sides and enclosed angle of the triangle in Fig. 1.11 are related by the cosine rule

$$u^2 = r^2 + R^2 - 2rR\cos\theta \qquad (1)$$

Instead of taking R outside the brackets as in (1.150), we can move r outside and write the expression for u as

$$\frac{1}{u} = \frac{1}{r}\left[1 - 2\left(\frac{R}{r}\right)\cos\theta + \left(\frac{R}{r}\right)^2\right]^{-1/2} \qquad (2)$$

Following the same treatment as in Section 1.12, but now with $h = R/r$ and $x = \cos\theta$, we get

$$\frac{1}{u} = \frac{1}{r}\left(1 - 2xh + h^2\right)^{-1/2} = \frac{1}{r}(1 - t)^{-1/2} \qquad (3)$$

where $t = 2xh - h^2$. The function $(1 - t)^{-1/2}$ is expanded as a binomial series, which again gives an infinite series in h, in which the coefficient of h^n is $P_n(x)$. The defining equation is as before:

$$\Psi(x, h) = \left(1 - 2xh + h^2\right)^{-1/2} = \sum_{n=0}^{\infty} h^n P_n(x) \qquad (4)$$

On substituting $h = R/r$ and $x = \cos\theta$, we find an alternative form for the generating equation for the Legendre polynomials:

$$\frac{1}{u} = \frac{1}{r}\sum_{n=0}^{\infty}\left(\frac{R}{r}\right)^n P_n(\cos\theta) \qquad (5)$$

On repeating the differentiation, and taking (1.159) into account, we get

$$\begin{aligned}
\frac{\partial^2}{\partial h^2}(h\Psi) &= \frac{\partial\Psi}{\partial h} + h(x - h)3\Psi^2\frac{\partial\Psi}{\partial h} + (x - 2h)\Psi^3 \\
&= (x - h)\Psi^3 + 3h(x - h)^2\Psi^5 + (x - 2h)\Psi^3 \\
&= (2x - 3h)\Psi^3 + 3h(x - h)^2\Psi^5
\end{aligned} \qquad (1.163)$$

Now substitute for Ψ^3, from (1.160), and Ψ^5, from (1.161), giving

$$\frac{\partial^2}{\partial h^2}(h\Psi) = (2x - 3h)\left(\frac{1}{h}\frac{\partial\Psi}{\partial x}\right) + 3h(x - h)^2\left(\frac{1}{3h^2}\frac{\partial^2\Psi}{\partial x^2}\right) \qquad (1.164)$$

Multiply throughout by h:

$$h\frac{\partial^2}{\partial h^2}(h\Psi) = (2x - 3h)\left(\frac{\partial\Psi}{\partial x}\right) + (x - h)^2\left(\frac{\partial^2\Psi}{\partial x^2}\right)$$

$$= 2x\left(\frac{\partial\Psi}{\partial x}\right) - 3h\left(\frac{\partial\Psi}{\partial x}\right) + (x - h)^2\left(\frac{\partial^2\Psi}{\partial x^2}\right) \qquad (1.165)$$

The second term on the right can be replaced as follows, again using (1.160) and (1.161):

$$3h\frac{\partial\Psi}{\partial x} = 3h^2\Psi^3 = \frac{1}{\Psi^2}\frac{\partial^2\Psi}{\partial x^2}$$

$$= \left(1 - 2xh + h^2\right)\frac{\partial^2\Psi}{\partial x^2} \qquad (1.166)$$

On substituting into (1.165) and gathering terms, we get

$$h\frac{\partial^2}{\partial h^2}(h\Psi) = \left[(x - h)^2 - (1 - 2xh + h^2)\right]\left(\frac{\partial^2\Psi}{\partial x^2}\right) + 2x\left(\frac{\partial\Psi}{\partial x}\right) \qquad (1.167)$$

$$h\frac{\partial^2}{\partial h^2}(h\Psi) = (x^2 - 1)\left(\frac{\partial^2\Psi}{\partial x^2}\right) + 2x\left(\frac{\partial\Psi}{\partial x}\right) \qquad (1.168)$$

The Legendre polynomials $P_n(x)$ are defined in (1.156) as the coefficients of h^n in the expansion of Ψ as a power series. On multiplying both sides of (1.156) by h, we get

$$h\Psi = \sum_{n=0}^{\infty} h^{n+1} P_n(x) \qquad (1.169)$$

We differentiate this expression twice and multiply by h to get a result that can be inserted on the left-hand side of (1.168):

$$\frac{\partial}{\partial h}(h\Psi) = \sum_{n=0}^{\infty} (n + 1)h^n P_n(x) \qquad (1.170)$$

$$h\frac{\partial^2}{\partial h^2}(h\Psi) = \sum_{n=0}^{\infty} n(n + 1)h^n P_n(x) \qquad (1.171)$$

Using (1.156), we can now eliminate Ψ and convert (1.168) into a second-order differential equation involving the Legendre polynomials $P_n(x)$,

Table 1.1. *Some ordinary Legendre polynomials of low degree*

n	$P_n(x)$	$P_n(\cos\theta)$
0	1	1
1	x	$\cos\theta$
2	$\frac{1}{2}(3x^2 - 1)$	$\frac{1}{2}(3\cos^2\theta - 1)$
3	$\frac{1}{2}(5x^3 - 3x)$	$\frac{1}{2}(5\cos^3\theta - 3\cos\theta)$
4	$\frac{1}{8}(35x^4 - 30x^2 + 3)$	$\frac{1}{8}(35\cos^4\theta - 30\cos^2\theta + 3)$

$$\sum_{n=0}^{\infty} h^n \left\{ (x^2 - 1)\frac{d^2 P_n(x)}{dx^2} + 2x\frac{dP_n(x)}{dx} \right\} = \sum_{n=0}^{\infty} n(n+1)h^n P_n(x) \quad (1.172)$$

$$\sum_{n=0}^{\infty} h^n \left\{ (x^2 - 1)\frac{d^2 P_n(x)}{dx^2} + 2x\frac{dP_n(x)}{dx} - n(n+1)P_n(x) \right\} = 0 \quad (1.173)$$

If this expression is true for every non-zero value of h, the quantity in curly brackets must be zero, thus

$$\left(1 - x^2\right)\frac{d^2 P_n(x)}{dx^2} - 2x\frac{dP_n(x)}{dx} + n(n+1)P_n(x) = 0 \quad (1.174)$$

An alternative, simpler form for this equation is obtained by combining the first two terms:

$$\frac{d}{dx}\left[\left(1 - x^2\right)\frac{dP_n(x)}{dx}\right] + n(n+1)P_n(x) = 0 \quad (1.175)$$

This is the *Legendre differential equation*. It has a family of solutions, each of which is a polynomial corresponding to a particular value of n. The *Legendre polynomials* provide solutions in potential analyses with spherical symmetry, and have an important role in geophysical theory. Some Legendre polynomials of low degree are listed in Table 1.1.

1.13.1 Orthogonality of the Legendre polynomials

Two vectors **a** and **b** are orthogonal if their scalar product is zero:

$$\mathbf{a} \cdot \mathbf{b} = a_x b_x + a_y b_y + a_z b_z = \sum_{i=1}^{3} a_i b_i = 0 \quad (1.176)$$

By analogy, two functions of the same variable are said to be orthogonal if their product, integrated over a particular range, is zero. For example, the trigonometric functions $\sin \theta$ and $\cos \theta$ are orthogonal for the range $0 \leq \theta \leq 2\pi$, because

$$\int_{\theta=0}^{2\pi} \sin \theta \cos \theta \, d\theta = \int_{\theta=0}^{2\pi} \frac{1}{2} \sin(2\theta) d\theta = -\frac{1}{4} \cos(2\theta) \Big|_{\theta=0}^{2\pi} = 0 \quad (1.177)$$

The Legendre polynomials $P_n(x)$ and $P_l(x)$ are orthogonal over the range $-1 \leq x \leq 1$. This can be established as follows. First, we write the Legendre equation in short form, dropping the variable x for both P_n and P_l, and, for brevity, writing

$$\frac{d}{dx} P_n(x) = P'_n \quad \text{and} \quad \frac{d^2}{dx^2} P_n(x) = P''_n \quad (1.178)$$

Thus

$$(1 - x^2) P''_n - 2x P'_n + n(n+1) P_n = 0 \quad (1.179)$$

$$(1 - x^2) P''_l - 2x P'_l + l(l+1) P_l = 0 \quad (1.180)$$

Multiplying (1.179) by P_l and (1.180) by P_n gives

$$(1 - x^2) P_l P''_n - 2x P_l P'_n + n(n+1) P_l P_n = 0 \quad (1.181)$$

$$(1 - x^2) P_n P''_l - 2x P'_l P_n + l(l+1) P_l P_n = 0 \quad (1.182)$$

Subtracting (1.182) from (1.181) gives

$$(1 - x^2)(P_l P''_n - P_n P''_l) - 2x(P_l P'_n - P'_l P_n) + [n(n+1) - l(l+1)] P_l P_n = 0$$

$$(1.183)$$

Note that

$$\frac{d}{dx}(P_l P'_n - P'_l P_n) = P_l P''_n + P'_l P'_n - P'_l P'_n - P''_l P_n = P_l P''_n - P''_l P_n$$

$$(1.184)$$

and

$$(1 - x^2) \frac{d}{dx}(P_l P'_n - P'_l P_n) - 2x(P_l P'_n - P'_l P_n)$$

$$= \frac{d}{dx}[(1 - x^2)(P_l P'_n - P'_l P_n)] \quad (1.185)$$

Thus

$$\frac{d}{dx}\left[(1-x^2)\left(P_l P'_n - P'_l P_n\right)\right] + [n(n+1) - l(l+1)]P_l P_n = 0 \quad (1.186)$$

Now integrate each term in this equation with respect to x over the range $-1 \le x \le 1$. We get

$$\left\{(1-x^2)\left(P_l P'_n - P'_l P_n\right)\right\}\big|^{+1}_{x=-1} + [n(n+1) - l(l+1)] \int\limits_{x=-1}^{+1} P_l P_n \, dx = 0$$
$$(1.187)$$

The first term is zero on evaluation of $(1-x^2)$ at $x = \pm 1$; thus the second term must also be zero. For $n \ne l$ the condition for orthogonality of the Legendre polynomials is

$$\int_{x=-1}^{+1} P_n(x) P_l(x) dx = 0 \quad (1.188)$$

1.13.2 Normalization of the Legendre polynomials

A function is said to be normalized if the integral of the square of the function over its range is equal to 1. Thus we must evaluate the integral $\int_{x=-1}^{+1} [P_n(x)]^2 \, dx$. We begin by recalling the generating function for the Legendre polynomials given in (1.156), which we rewrite for $P_n(x)$ and $P_l(x)$ individually:

$$\sum_{n=0}^{\infty} h^n P_n(x) = \left(1 - 2xh + h^2\right)^{-1/2} \quad (1.189)$$

$$\sum_{l=0}^{\infty} h^l P_l(x) = \left(1 - 2xh + h^2\right)^{-1/2} \quad (1.190)$$

Multiplying these equations together gives

$$\sum_{l=0}^{\infty}\sum_{n=0}^{\infty} h^{n+l} P_n(x) P_l(x) = \left(1 - 2xh + h^2\right)^{-1} \quad (1.191)$$

Now let $l = n$ and integrate both sides with respect to x, taking into account (1.188):

$$\sum_{n=0}^{\infty} h^{2n} \int\limits_{x=-1}^{+1} [P_n(x)]^2 \, dx = \int\limits_{x=-1}^{+1} \frac{dx}{1 + h^2 - 2xh} \quad (1.192)$$

The right-hand side of this equation is a standard integration that results in a natural logarithm:

$$\int \frac{dx}{a+bx} = \frac{1}{b}\ln(a+bx) \tag{1.193}$$

The right-hand side of (1.192) therefore leads to

$$\int_{x=-1}^{+1} \frac{dx}{1+h^2-2xh} = \frac{1}{(-2h)}\ln(1+h^2-2xh)\Big|_{x=-1}^{+1}$$

$$= \frac{-1}{2h}\left[\ln(1+h^2-2h)-\ln(1+h^2+2h)\right] \tag{1.194}$$

and

$$\int_{x=-1}^{+1} \frac{dx}{1+h^2-2xh} = \frac{1}{h}\left[-\frac{1}{2}\ln(1-h)^2+\frac{1}{2}\ln(1+h)^2\right]$$

$$= \frac{1}{h}[\ln(1+h)-\ln(1-h)] \tag{1.195}$$

Using the MacLaurin series for the natural logarithms as in (1.135), we get

$$\ln(1+h) = h - \frac{h^2}{2} + \frac{h^3}{3} - \frac{h^4}{4} + \cdots (-1)^{n-1}\frac{h^n}{n} + \cdots \tag{1.196}$$

$$\ln(1-h) = -h - \frac{h^2}{2} - \frac{h^3}{3} - \frac{h^4}{4} - \cdots (-1)^{n-1}\frac{(-h)^n}{n} + \cdots \tag{1.197}$$

Subtracting the second equation from the first gives

$$\int_{x=-1}^{+1} \frac{dx}{1+h^2-2xh} = \frac{2}{h}\left[h+\frac{h^3}{3}+\frac{h^5}{5}+\cdots\right] = \frac{2}{h}\sum_{n=0}^{\infty}\frac{h^{2n+1}}{2n+1} \tag{1.198}$$

Inserting this result into (1.192) gives

$$\sum_{n=0}^{\infty} h^{2n}\int_{x=-1}^{+1}[P_n(x)]^2\,dx = \frac{2}{h}\sum_{n=0}^{\infty}\frac{h^{2n+1}}{2n+1} \tag{1.199}$$

$$\sum_{n=0}^{\infty} h^{2n}\left(\int_{x=-1}^{+1}[P_n(x)]^2\,dx - \frac{2}{2n+1}\right) = 0 \tag{1.200}$$

This is true for every value of h in the summation, so we obtain the normalizing condition for the Legendre polynomials:

$$\int_{x=-1}^{+1} [P_n(x)]^2 \, dx = \frac{2}{2n+1} \qquad (1.201)$$

It follows that $\left(n + \frac{1}{2}\right)^{1/2} P_n(x)$ is a *normalized* Legendre polynomial.

1.14 Rodrigues' formula

The Legendre polynomials can be easily computed with the aid of a formula derived by a French mathematician, Olinde Rodrigues (1795–1851). First, we define the function

$$f(x) = \left(x^2 - 1\right)^n \qquad (1.202)$$

Differentiating $f(x)$ once with respect to x gives

$$\frac{df}{dx} = \frac{d}{dx}\left(x^2 - 1\right)^n = 2nx\left(x^2 - 1\right)^{n-1} \qquad (1.203)$$

Multiplying the result by $(x^2 - 1)$ gives

$$\left(x^2 - 1\right) \frac{d}{dx}\left(x^2 - 1\right)^n = 2nx\left(x^2 - 1\right)^n \qquad (1.204)$$

$$\left(x^2 - 1\right) \frac{df}{dx} = 2nxf \qquad (1.205)$$

Now we use Leibniz's rule (1.144) to differentiate both sides of this equation $n + 1$ times with respect to x. Writing $D = d/dx$ as in Section 1.11,

$$D^{n+1}(uv) = \sum_{k=0}^{n+1} \frac{(n+1)!}{k!(n+1-k)!}\left(D^k u\right)\left(D^{n+1-k}v\right) \qquad (1.206)$$

On the left-hand side of (1.205) let $u(x) = (x^2 - 1)$ and $v(x) = df/dx = Df$. Applying Leibniz's rule, we note that after only three differentiations of $(x^2 - 1)$ the result is zero and the series is curtailed.

On the right-hand side let $u(x) = 2nx$ and $v(x) = f$. Note that in this case the series is curtailed after two differentiations.

Thus, using Leibniz's rule to differentiate each side of (1.205) $n + 1$ times, we get

$$(x^2 - 1)D^{n+2}f + 2x(n+1)D^{n+1}f + 2\frac{(n+1)n}{1\cdot2}D^nf = 2nx\,D^{n+1}f$$
$$+ 2n(n+1)D^nf$$
$$(1.207)$$

On gathering terms and bringing all to the left-hand side, we have

$$(x^2 - 1)D^{n+2}f + 2x\,D^{n+1}f - n(n+1)D^nf = 0 \qquad (1.208)$$

Now we define $y(x)$ such that

$$y(x) = D^nf = \frac{d^n}{dx^n}(x^2 - 1)^n \qquad (1.209)$$

and we have

$$(x^2 - 1)\frac{d^2y}{dx^2} + 2x\frac{dy}{dx} - n(n+1)y = 0 \qquad (1.210)$$

On comparing with (1.174), we see that this is the Legendre equation. The Legendre polynomials must therefore be proportional to $y(x)$, so we can write

$$P_n(x) = c_n\frac{d^n}{dx^n}(x^2 - 1)^n \qquad (1.211)$$

The quantity c_n is a calibration constant. To determine c_n we first write

$$\frac{d^n}{dx^n}(x^2 - 1)^n = \frac{d^n}{dx^n}[(x-1)^n(x+1)^n] \qquad (1.212)$$

then we apply Leibniz's rule to the product on the right-hand side of the equation:

$$\frac{d^n}{dx^n}(x^2 - 1)^n = \sum_{m=0}^{n}\frac{n!}{m!(n-m)!}\frac{d^m}{dx^m}(x-1)^n\frac{d^{n-m}}{dx^{n-m}}(x+1)^n \qquad (1.213)$$

The successive differentiations of $(x-1)^n$ give

$$\frac{d}{dx}(x-1)^n = n(x-1)^{n-1}$$
$$\frac{d^2}{dx^2}(x-1)^n = n(n-1)(x-1)^{n-2}$$
$$\frac{d^{n-1}}{dx^{n-1}}(x-1)^n = (n(n-1)(n-2)\ldots3\cdot2\cdot1)(x-1) = n!(x-1)$$
$$\frac{d^n}{dx^n}(x-1)^n = n!$$
$$(1.214)$$

Each differentiation in (1.214) is zero at $x = 1$, except the last one. Thus each term in the sum in (1.213) is also zero except for the last one, for which $m = n$. Substituting $x = 1$ gives

$$\left[\frac{d^n}{dx^n}(x^2 - 1)^n\right]_{x=1} = \left[(x+1)^n\frac{d^n}{dx^n}(x-1)^n\right]_{x=1} = 2^n n! \qquad (1.215)$$

Putting this result and the condition $P_n(1) = 1$ into (1.211) gives

$$P_n(1) = c_n\left(\frac{d^n}{dx^n}(x^2 - 1)^n\right)_{x=1} = c_n 2^n n! = 1 \qquad (1.216)$$

where

$$c_n = \frac{1}{2^n n!} \qquad (1.217)$$

Rodrigues' formula for the Legendre polynomials is therefore

$$P_n(x) = \frac{1}{2^n n!}\frac{d^n}{dx^n}(x^2 - 1)^n \qquad (1.218)$$

1.15 Associated Legendre polynomials

Many physical properties of the Earth, such as its magnetic field, are not azimuthally symmetric about the rotation axis when examined in detail. However, these properties can be described using mathematical functions that are based upon the Legendre polynomials described in the preceding section. To derive these functions, we start from the Legendre equation, (1.174), which can be written in shorthand form as

$$(1 - x^2)P_n'' - 2xP_n' + n(n+1)P_n = 0 \qquad (1.219)$$

Now we differentiate this equation with respect to x:

$$(1 - x^2)\frac{d}{dx}P_n'' - 2xP_n'' - 2x\frac{d}{dx}P_n' - 2P_n' + n(n+1)\frac{d}{dx}P_n = 0 \quad (1.220)$$

On noting that we can equally write $P_n'' = (d/dx)P_n'$ and $P_n' = (d/dx)P_n$, this can be written alternatively as

$$(1 - x^2)\frac{d}{dx}P_n'' - 4x\frac{d}{dx}P_n' + [n(n+1) - 2]\frac{d}{dx}P_n = 0 \qquad (1.221)$$

which can be written, for later comparison,

$$(1 - x^2)\frac{d}{dx}P_n'' - 2(2)x\frac{d}{dx}P_n' + [n(n+1) - 1(2)]\frac{d}{dx}P_n = 0 \quad (1.222)$$

Next, we differentiate this expression again, observing the same rules and gathering terms,

$$\left((1-x^2)\frac{d^2}{dx^2}P_n'' - 2x\frac{d}{dx}P_n''\right) - \left(4\frac{d}{dx}P_n' + 4x\frac{d^2}{dx^2}P_n'\right) + [n(n+1) - 2]\frac{d^2}{dx^2}P_n = 0$$
$$(1.223)$$

$$(1 - x^2)\frac{d^2}{dx^2}P_n'' - 2x\frac{d^2}{dx^2}P_n' - 4x\frac{d^2}{dx^2}P_n' + [n(n+1) - 2 - 4]\frac{d^2}{dx^2}P_n = 0$$
$$(1.224)$$

$$(1 - x^2)\frac{d^2}{dx^2}P_n'' - 6x\frac{d^2}{dx^2}P_n' + [n(n+1) - 6]\frac{d^2}{dx^2}P_n = 0 \quad (1.225)$$

which, as we did with (1.222), we can write in the form

$$(1 - x^2)\frac{d^2}{dx^2}P_n'' - 2(3)x\frac{d^2}{dx^2}P_n' + [n(n+1) - 2(3)]\frac{d^2}{dx^2}P_n = 0 \quad (1.226)$$

On following step-by-step in the same manner, we get after the third differentiation

$$(1 - x^2)\frac{d^3}{dx^3}P_n'' - 2(4)x\frac{d^3}{dx^3}P_n' + [n(n+1) - 3(4)]\frac{d^3}{dx^3}P_n = 0 \quad (1.227)$$

Equations (1.222), (1.226), and (1.227) all have the same form. The higher-order differentiation is accompanied by systematically different constants. By extension, differentiating (1.219) m times (where $m \le n$) yields the differential equation

$$(1 - x^2)\frac{d^m}{dx^m}P_n'' - 2(m+1)x\frac{d^m}{dx^m}P_n' + [n(n+1) - m(m+1)]\frac{d^m}{dx^m}P_n = 0$$
$$(1.228)$$

Now let the mth-order differentiation of P_n be written as

$$\frac{d^m}{dx^m}P_n(x) = \frac{Q(x)}{(1 - x^2)^{m/2}} \quad (1.229)$$

Substitution of this expression into (1.228) gives a new differential equation involving $Q(x)$. We need to determine both $(d^m/dx^m)P_n'$ and $(d^m/dx^m)P_n''$, so first we differentiate (1.229) with respect to x:

$$\frac{d^m}{dx^m}P'_n = \frac{Q'}{(1-x^2)^{m/2}} - \left(\frac{m}{2}\right)(-2x)\frac{Q}{(1-x^2)^{m/2+1}} \tag{1.230}$$

$$\frac{d^m}{dx^m}P'_n = (1-x^2)^{-(m+2)/2}\{(1-x^2)Q' + mxQ\} \tag{1.231}$$

A further differentiation of (1.231) by parts gives

$$\frac{d^m}{dx^m}P''_n = \left(\frac{d}{dx}(1-x^2)^{-(m+2)/2}\right)\{(1-x^2)Q' + mxQ\}$$
$$+ (1-x^2)^{-(m+2)/2}\frac{d}{dx}\{(1-x^2)Q' + mxQ\}$$
$$= (m+2)x(1-x^2)^{-(m+2)/2-1}\{(1-x^2)Q' + mxQ\}$$
$$+ (1-x^2)^{-(m+2)/2}\{(1-x^2)Q'' + mxQ' - 2xQ' + mQ\} \tag{1.232}$$

$$\frac{d^m}{dx^m}P''_n = (1-x^2)^{-(m+2)/2}\Big\{(1-x^2)Q'' + (m-2)xQ' + mQ + (m+2)xQ'$$
$$+ \frac{m(m+2)x^2Q}{1-x^2}\Big\} \tag{1.233}$$

$$\frac{d^m}{dx^m}P''_n = (1-x^2)^{-(m+2)/2}\Big\{(1-x^2)Q'' + 2mxQ' + mQ + \frac{m(m+2)x^2Q}{1-x^2}\Big\} \tag{1.234}$$

Now we substitute (1.231) and (1.234) into (1.228). Unless the multiplier $(1-x^2)^{-(m+2)/2}$ is always zero, Q must satisfy the following equation:

$$(1-x^2)^2Q'' + 2mx(1-x^2)Q' + m(1-x^2)Q + m(m+2)x^2Q$$
$$- 2(m+1)x(1-x^2)Q' - 2m(m+1)x^2Q$$
$$+ [n(n+1) - m(m+1)](1-x^2)Q = 0 \tag{1.235}$$

The remainder of the evaluation consists of gathering and reducing terms; we finally get

$$(1-x^2)Q'' - 2xQ' + \left[n(n+1) - \frac{m^2}{1-x^2}\right]Q = 0 \tag{1.236}$$

The functions $Q(x)$ involve two parameters, the degree n and order m, and are written $P_{n,m}(x)$. Thus

$$\left(1 - x^2\right)\frac{d^2}{dx^2}P_{n,m}(x) - 2x\frac{d}{dx}P_{n,m}(x) + \left[n(n+1) - \frac{m^2}{1-x^2}\right]P_{n,m}(x) = 0$$

$$(1.237)$$

This is the associated Legendre equation. The solutions $P_{n,m}(x)$ or $P_{n,m}(\cos\theta)$, where $x = \cos\theta$, are called *associated Legendre polynomials*, and are obtained from the ordinary Legendre polynomials using the definition of Q in (1.229):

$$P_{n,m}(x) = \left(1 - x^2\right)^{m/2}\frac{d^m}{dx^m}P_n(x) \qquad (1.238)$$

Substituting Rodrigues' formula (1.218) for $P_n(x)$ into this equation gives

$$P_{n,m}(x) = \frac{\left(1 - x^2\right)^{m/2}}{2^n n!}\frac{d^{n+m}}{dx^{n+m}}\left(x^2 - 1\right)^n \qquad (1.239)$$

The highest power of x in the function $(x^2 - 1)^n$ is x^{2n}. After $2n$ differentiations the result will be a constant, and a further differentiation will give zero. Therefore $n + m \leq 2n$, and possible values of m are limited to the range $0 \leq m \leq n$.

1.15.1 Orthogonality of associated Legendre polynomials

For succinctness we again write $P_{n,m}(x)$ as simply $P_{n,m}$. The defining equations for the associated Legendre polynomials $P_{n,m}$ and $P_{l,m}$ are

$$\left(1 - x^2\right)\left(P_{n,m}\right)'' - 2x\left(P_{n,m}\right)' + \left[n(n+1) - \frac{m^2}{1-x^2}\right]P_{n,m} = 0 \qquad (1.240)$$

$$\left(1 - x^2\right)\left(P_{l,m}\right)'' - 2x\left(P_{l,m}\right)' + \left[l(l+1) - \frac{m^2}{1-x^2}\right]P_{l,m} = 0 \qquad (1.241)$$

As for the ordinary Legendre polynomials, we multiply (1.240) by $P_{l,m}$ and (1.241) by $P_{n,m}$:

$$\left(1 - x^2\right)\left(P_{n,m}\right)''P_{l,m} - 2x\left(P_{n,m}\right)'P_{l,m} + \left[n(n+1) - \frac{m^2}{1-x^2}\right]P_{n,m}P_{l,m} = 0$$

$$(1.242)$$

$$\left(1 - x^2\right)\left(P_{l,m}\right)'' P_{n,m} - 2x\left(P_{l,m}\right)' P_{n,m} + \left[l(l+1) - \frac{m^2}{1 - x^2}\right] P_{n,m} P_{l,m} = 0 \tag{1.243}$$

On subtracting (1.243) from (1.242) we have

$$\left(1 - x^2\right)\left[\left(P_{n,m}\right)'' P_{l,m} - \left(P_{l,m}\right)'' P_{n,m}\right] - 2x\left[\left(P_{n,m}\right)' P_{l,m} - \left(P_{l,m}\right)' P_{n,m}\right]$$
$$+ \left[n(n+1) - l(l+1)\right] P_{n,m} P_{l,m} = 0 \tag{1.244}$$

Following the method used to establish the orthogonality of the ordinary Legendre polynomials (Section 1.13.1), we can write this equation as

$$\frac{d}{dx}\left\{\left(1 - x^2\right)\left(\left(P_{n,m}\right)' P_{l,m} - \left(P_{l,m}\right)' P_{n,m}\right)\right\} + \left[n(n+1) - l(l+1)\right] P_{n,m} P_{l,m}$$
$$= 0 \tag{1.245}$$

On integrating each term with respect to x over the range $-1 \leq x \leq 1$, we get

$$\left\{\left(1 - x^2\right)\left(\left(P_{n,m}\right)' P_{l,m} - \left(P_{l,m}\right)' P_{n,m}\right)\right\}\Big|_{x=-1}^{+1}$$
$$+ \left[n(n+1) - l(l+1)\right] \int_{x=-1}^{+1} P_{n,m} P_{l,m}\, dx = 0 \tag{1.246}$$

The first term is zero on evaluation of $(1 - x^2)$ at $x = \pm 1$; thus the second term must also be zero. Provided that $n \neq l$, the condition of orthogonality of the associated Legendre polynomials is

$$\int_{x=-1}^{x=+1} P_{n,m}(x) P_{l,m}(x)\, dx = 0 \tag{1.247}$$

1.15.2 Normalization of associated Legendre polynomials

Squaring the associated Legendre polynomials and integrating over $-1 \leq x \leq 1$ gives

$$\int_{x=-1}^{x=+1} \left[P_{n,m}(x)\right]^2 dx = \frac{2}{2n+1} \frac{(n+m)!}{(n-m)!} \tag{1.248}$$

The squared functions do not integrate to 1, so they are not normalized. If each polynomial is multiplied by a normalizing function, the integrated squared polynomial can be made to equal a chosen value. Different conditions for this apply in geodesy and geomagnetism.

The Legendre polynomials used in *geodesy* are *fully normalized*. They are defined as follows:

$$P_n^m(x) = \left(\left(\frac{2n+1}{2} \right) \frac{(n-m)!}{(n+m)!} \right)^{1/2} P_{n,m}(x) \tag{1.249}$$

The Legendre polynomials used in *geomagnetism* are *partially normalized* (or *quasi-normalized*). Schmidt in 1889 defined this method of normalization so that

$$P_n^m(x) = \left(2 \frac{(n-m)!}{(n+m)!} \right)^{1/2} P_{n,m}(x), \qquad m \neq 0 \tag{1.250}$$

$$P_n^0(x) = P_{n,0}(x), \qquad m = 0 \tag{1.251}$$

Integration of the squared Schmidt polynomials over the full range $-1 \leq x \leq 1$ gives the value 1 for $m = 0$ and $1/(2n+1)$ for $m > 0$.

Some fully normalized Legendre polynomials and partially normalized Schmidt polynomials are listed in Table 1.2.

Table 1.2. *Some fully normalized associated Legendre polynomials and partially normalized Schmidt polynomials of low degree and order*

n	m	$P_n^m(\cos\theta)$, Legendre, fully normalized	$P_n^m(\cos\theta)$, Schmidt, partially normalized
1	0	$\cos\theta$	$\cos\theta$
1	1	$\sin\theta$	$\sin\theta$
2	0	$\frac{1}{2}(3\cos^2\theta - 1)$	$\frac{1}{2}(3\cos^2\theta - 1)$
2	1	$3\sin\theta\cos\theta$	$\sqrt{3}\sin\theta\cos\theta$
2	2	$3\sin^2\theta$	$\frac{\sqrt{3}}{2}\sin^2\theta$
3	0	$\frac{1}{2}\cos\theta(5\cos^2\theta - 3)$	$\frac{1}{2}\cos\theta(5\cos^2\theta - 3)$
3	1	$\frac{3}{2}\sin\theta(5\cos^2\theta - 1)$	$\frac{\sqrt{6}}{4}\sin\theta(5\cos^2\theta - 1)$
3	2	$15\sin^2\theta\cos\theta$	$\frac{\sqrt{15}}{2}15\sin^2\theta\cos\theta$
3	3	$15\sin^3\theta$	$\frac{\sqrt{10}}{4}\sin^3\theta$

1.16 Spherical harmonic functions

Several geophysical potential fields – for example, gravitation and geomagnetism – satisfy the Laplace equation. Spherical polar coordinates are best suited for describing a global geophysical potential. The potential can vary with distance r from the Earth's center and with polar angular distance θ and azimuth ϕ (equivalent to co-latitude and longitude in geographic terms) on any concentric spherical surface. The solution of Laplace's equation in spherical polar coordinates for a potential U may be written (see Section 2.4.5, (2.104))

$$U = \sum_{n=0}^{\infty} \sum_{m=0}^{n} \left(A_n r^n + \frac{B_n}{r^{n+1}} \right) (a_n^m \cos(m\phi) + b_n^m \sin(m\phi)) P_n^m(\cos\theta) \quad (1.252)$$

Here A_n, B_n, a_n^m, and b_n^m are constants that apply to a particular situation. On the surface of the Earth, or an arbitrary sphere, the radial part of the potential of a point source at the center of the sphere has a constant value and the variation over the surface of the sphere is described by the functions in θ and ϕ. We are primarily interested in solutions outside the Earth, for which A_n is zero. Also we can set the constant B_n equal to R^{n+1}, where R is the Earth's mean radius. The potential is then given by

$$U = \sum_{n=0}^{\infty} \sum_{m=0}^{n} \left(\frac{R}{r} \right)^{n+1} (a_n^m \cos(m\phi) + b_n^m \sin(m\phi)) P_n^m(\cos\theta) \quad (1.253)$$

Let the *spherical harmonic functions* $C_n^m(\theta, \phi)$ and $S_n^m(\theta, \phi)$ be defined as

$$\begin{aligned} C_n^m(\theta, \phi) &= \cos(m\phi) \cdot P_n^m(\cos\theta) \\ S_n^m(\theta, \phi) &= \sin(m\phi) \cdot P_n^m(\cos\theta) \end{aligned} \quad (1.254)$$

The variation of the potential over the surface of a sphere may be described by these functions, or a more general spherical harmonic function $Y_n^m(\theta, \phi)$ that combines the sine and cosine variations:

$$Y_n^m(\theta, \phi) = P_n^m(\cos\theta) \begin{Bmatrix} \cos(m\phi) \\ \sin(m\phi) \end{Bmatrix} \quad (1.255)$$

Like their constituent parts – the sine, cosine, and associated Legendre functions – spherical harmonic functions are orthogonal and can be normalized.

1.16.1 Normalization of spherical harmonic functions

Normalization of the functions $C_n^m(\theta, \phi)$ and $S_n^m(\theta, \phi)$ requires integrating the squared value of each function over the surface of a unit sphere. The element of surface area on a unit sphere is $d\Omega = \sin\theta\, d\theta\, d\phi$ (Box 1.3) and the limits of integration are $0 \leq \theta \leq \pi$ and $0 \leq \phi \leq 2\pi$. The integral is

$$\iint_S (C_n^m(\theta, \phi))^2\, d\Omega = \int_{\theta=0}^{\pi} \int_{\phi=0}^{2\pi} (C_n^m(\theta, \phi))^2 \sin\theta\, d\theta\, d\phi$$

$$= \int_{\theta=0}^{\pi} \int_{\phi=0}^{2\pi} (\cos(m\phi) P_n^m(\cos\theta))^2 \sin\theta\, d\theta\, d\phi \quad (1.256)$$

Let $x = \cos\theta$ in the associated Legendre polynomial, so that $dx = -\sin\theta\, d\theta$ and the limits of integration are $-1 \leq x \leq 1$. The integration becomes

$$\int_{x=-1}^{1} \left\{ \int_{\phi=0}^{2\pi} \cos^2(m\phi) d\phi \right\} [P_n^m(x)]^2\, dx = \pi \int_{x=-1}^{+1} [P_n^m(x)]^2\, dx \quad (1.257)$$

Normalization of the associated Legendre polynomials gives the result in (1.248), thus

$$\iint_S (C_n^m(\theta, \phi))^2\, d\Omega = \left(\frac{2\pi}{2n+1} \right) \frac{(n+m)!}{(n-m)!} \quad (1.258)$$

The normalization of the function $S_n^m(\theta, \phi)$ by this method delivers the same result.

Spherical harmonic functions make it possible to express the variation of a physical property (e.g., gravity anomalies, $g(\theta, \phi)$) on the surface of the Earth as an infinite series, such as

$$g(\theta, \phi) = \sum_{n=0}^{\infty} \sum_{m=0}^{n} (a_n^m C_n^m(\theta, \phi) + b_n^m S_n^m(\theta, \phi)) \quad (1.259)$$

The coefficients a_n^m and b_n^m may be obtained by multiplying the function $g(\theta, \phi)$ by $C_n^m(\theta, \phi)$ or $S_n^m(\theta, \phi)$, respectively, and integrating the product over the surface of the unit sphere. The normalization properties give

$$a_n^m = \left(\frac{2n+1}{2\pi}\right)\frac{(n-m)!}{(n+m)!}\iint\limits_S g(\theta,\phi)\cdot C_n^m(\theta,\phi)d\Omega$$

$$b_n^m = \left(\frac{2n+1}{2\pi}\right)\frac{(n-m)!}{(n+m)!}\iint\limits_S g(\theta,\phi)\cdot S_n^m(\theta,\phi)d\Omega$$

$$(1.260)$$

1.16.2 Zonal, sectorial, and tesseral spherical harmonics

The spherical harmonic functions $Y_n^m(\theta,\phi)$ have geometries that allow graphic representation of a potential on the surface of a sphere. Deviations of the potential from a constant value form alternating regions in which the potential is larger or smaller than a uniform value. Where the potential surface intersects the spherical surface a nodal line is formed. The appearance of any $Y_n^m(\theta,\phi)$ is determined by the distribution of its nodal lines. These occur where $Y_n^m(\theta,\phi)=0$. To simplify the discussion we will associate a constant value of the polar angle θ with a circle of latitude, and a constant value of the azimuthal angle ϕ with a circle of longitude.

The definition of the associated Legendre polynomials in (1.239) shows that the equation $P_n^m(x)=0$ has $n-m$ roots, apart from the trivial solution $x=\pm1$. The variation of the spherical harmonic $Y_n^m(\theta,\phi)$ with latitude θ thus has $n-m$ nodal lines, each a circle of latitude, between the two poles. If additionally $m=0$, the potential on the sphere varies only with latitude and there are n nodal lines separating *zones* in which the potential is greater or less than the uniform value. An example of a *zonal* spherical harmonic is $Y_2^0(\theta,\phi)$, shown in Fig. 1.12(a).

The solution of Laplace's equation (1.253) shows that the variation in potential around any circle of latitude is described by the function

$$\Phi(\phi) = a_n^m\cos(m\phi) + b_n^m\sin(m\phi) \tag{1.261}$$

There are $2m$ nodal lines where $\Phi(\phi)=0$, corresponding to $2m$ meridians of longitude, or m great circles. In the special case in which $n=m$, there are no

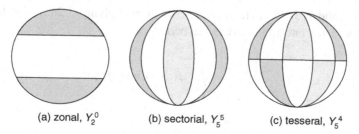

(a) zonal, Y_2^0 (b) sectorial, Y_5^5 (c) tesseral, Y_5^4

Fig. 1.12. Appearance of (a) zonal, (b) sectorial, and (c) tesseral spherical harmonics, projected on a meridian plane of the reference sphere.

nodal lines of latitude and the longitudinal lines separate *sectors* in which the potential is greater or less than the uniform value. An example of a *sectorial* spherical harmonic is $Y_5^5(\theta, \phi)$, shown in Fig. 1.12(b).

In the general case ($m \neq 0$, $n \neq m$) the potential varies with both latitude and longitude. There are $n - m$ nodal lines of latitude and m nodal great circles ($2m$ meridians) of longitude. The appearance of the spherical harmonic resembles a patchwork of alternating regions in which the potential is greater or less than the uniform value. An example of a *tesseral* spherical harmonic is $Y_5^4(\theta, \phi)$, which is shown in Fig. 1.12(c).

1.17 Fourier series, Fourier integrals, and Fourier transforms

1.17.1 Fourier series

Analogously to the representation of a continuous function by a power series (Section 1.10), it is possible to represent a periodic function by an infinite sum of terms consisting of the sines and cosines of harmonics of a fundamental frequency. Consider a periodic function $f(t)$ with period τ that is defined in the interval $0 \leq t \leq \tau$, so that (a) $f(t)$ is finite within the interval; (b) $f(t)$ is periodic outside the interval, i.e., $f(t + \tau) = f(t)$; and (c) $f(t)$ is single-valued in the interval except at a finite number of points, and is continuous between these points. Conditions (a)–(c) are known as the *Dirichlet conditions*. If they are satisfied, $f(t)$ can be represented as

$$f(t) = \frac{a_0}{2} + \sum_{n=1}^{\infty} (a_n \cos(n\omega t) + b_n \sin(n\omega t)) \qquad (1.262)$$

where $\omega = 2\pi/\tau$ and the factor $\frac{1}{2}$ in the first term is included for reasons of symmetry. This representation of $f(t)$ is known as a *Fourier series*. The orthogonal properties of sine and cosine functions allow us to find the coefficients a_n and b_n of the nth term in the series by multiplying (1.262) by $\sin(n\omega t)$ or $\cos(n\omega t)$ and integrating over a full period:

$$a_n = \frac{2}{\tau} \int_{t=-\tau/2}^{\tau/2} f(t)\cos(n\omega t)dt$$

$$\qquad (1.263)$$

$$b_n = \frac{2}{\tau} \int_{t=-\tau/2}^{\tau/2} f(t)\sin(n\omega t)dt$$

Instead of using trigonometric functions, we can replace the sine and cosine terms with complex exponentials using the definitions in (1.7), i.e., we write

$$
\begin{aligned}
\cos(n\omega t) &= \frac{\exp(in\omega t) + \exp(-in\omega t)}{2} \\
\sin(n\omega t) &= \frac{\exp(in\omega t) - \exp(-in\omega t)}{2i}
\end{aligned}
\tag{1.264}
$$

Using these relationships in (1.262) yields

$$
\begin{aligned}
f(t) &= \sum_{n=0}^{\infty} \left(\frac{a_n}{2} [\exp(in\omega t) + \exp(-in\omega t)] + \frac{b_n}{2i} [\exp(in\omega t) - \exp(-in\omega t)] \right) \\
&= \sum_{n=0}^{\infty} \left(\frac{a_n - ib_n}{2} \exp(in\omega t) \right) + \sum_{n=0}^{\infty} \left(\frac{a_n + ib_n}{2} \exp(-in\omega t) \right)
\end{aligned}
\tag{1.265}
$$

The summation indices are dummy variables, so in the second sum we can replace n by $-n$, and extend the limits of the sum to $n = -\infty$; thus

$$
\begin{aligned}
f(t) &= \sum_{n=0}^{\infty} \left(\frac{a_n - ib_n}{2} \exp(in\omega t) \right) + \sum_{n=-\infty}^{0} \left(\frac{a_{-n} + ib_{-n}}{2} \exp(in\omega t) \right) \\
&= \sum_{n=-\infty}^{\infty} \frac{(a_n + a_{-n}) - i(b_n - b_{-n})}{2} \exp(in\omega t)
\end{aligned}
\tag{1.266}
$$

If we define c_n as the complex number

$$
c_n = \frac{(a_n + a_{-n}) - i(b_n - b_{-n})}{2}
\tag{1.267}
$$

the Fourier series (1.262) can be written in complex exponential form as

$$
f(t) = \sum_{n=-\infty}^{\infty} c_n \exp(in\omega t)
\tag{1.268}
$$

In this case the harmonic coefficients c_n are given by

$$
c_n = \frac{1}{\tau} \int_{t=-\tau/2}^{\tau/2} f(t) \exp(-in\omega t) dt
\tag{1.269}
$$

1.17.2 Fourier integrals and Fourier transforms

A Fourier series represents the periodic behavior of a physical property as an infinite set of discrete frequencies. The theory can be extended to represent a function $f(t)$ that is not periodic and is made up of a continuous spectrum of frequencies, provided that the function satisfies the Dirichlet conditions specified above and that it has a finite energy:

$$\int_{t=-\infty}^{\infty} |f(t)|^2 \, dt < \infty \tag{1.270}$$

The infinite sum in (1.268) is replaced by a *Fourier integral* and the complex coefficients c_n are replaced by an amplitude function $g(\omega)$:

$$f(t) = \int_{\omega=-\infty}^{\infty} g(\omega)\exp(i\omega t)d\omega \tag{1.271}$$

where $g(\omega)$ is a continuous function, obtained from the equation

$$g(\omega) = \frac{1}{2\pi} \int_{t=-\infty}^{\infty} f(t)\exp(-i\omega t)dt \tag{1.272}$$

The transition from Fourier series to Fourier integral is explained in Box 1.5. The function $g(\omega)$ is called the *forward Fourier transform* of $f(t)$, and $f(t)$ is called the *inverse Fourier transform* of $g(\omega)$. Fourier transforms constitute a powerful mathematical tool for transforming a function $f(t)$ that is known in the time domain into a new function $g(\omega)$ in the frequency domain.

1.17.3 Fourier sine and cosine transforms

A simple but important characteristic of a function is whether it is even or odd. An even function has the same value for both positive and negative values of its argument, i.e., $f(-t) = f(t)$. The cosine of an angle is an example of an even function. The integral of an even function over a symmetric interval about the origin is equal to twice the integral of the function over the positive argument. The sign of an odd function changes with that of the argument, i.e., $f(-t) = -f(t)$. For example, the sine of an angle is an odd function. The integral of an odd function over a symmetric interval about the origin is zero. The product of two odd functions or two even functions is an even function; the product of an odd function and an even function is an odd function.

Box 1.5. Transition from Fourier series to Fourier integral

The complex exponential Fourier series for a function $f(t)$ is

$$f(t) = \sum_{n=-\infty}^{\infty} c_n \exp(in\omega t) \tag{1}$$

where the complex coefficients c_n are given by

$$c_n = \frac{1}{\tau} \int_{t=-\tau/2}^{\tau/2} f(t)\exp(-in\omega t)dt \tag{2}$$

In these expressions $\omega = 2\pi/\tau$ is the fundamental frequency and τ is the fundamental period. From one value of n to the next, the harmonic frequency changes by $\delta\omega = 2\pi/\tau$, so the factor preceding the second equation can be replaced by $1/\tau = \delta\omega/(2\pi)$. To avoid confusion when we insert (2) into (1), we change the dummy variable of the integration to u, giving

$$c_n = \frac{\delta\omega}{2\pi} \int_{u=-\tau/2}^{\tau/2} f(u)\exp(-in\omega u)du \tag{3}$$

After insertion, (1) becomes

$$f(t) = \sum_{n=-\infty}^{\infty} \left(\frac{\delta\omega}{2\pi} \int_{u=-\tau/2}^{\tau/2} f(u)\exp(-in\omega u)du \right) \exp(in\omega t)$$

$$= \sum_{n=-\infty}^{\infty} \left(\frac{\delta\omega}{2\pi} \int_{u=-\tau/2}^{\tau/2} f(u)\exp(in\omega(t-u))du \right) \tag{4}$$

We now define the function within the integral as

$$F(\omega) = \int_{u=-\tau/2}^{\tau/2} f(u)\exp(in\omega(t-u))du \tag{5}$$

The initial Fourier series becomes

$$f(t) = \frac{1}{2\pi} \sum_{\omega=-\infty}^{\infty} F(\omega)\delta\omega \tag{6}$$

We now let the incremental frequency $\delta\omega$ become very small, tending in the limit to zero; this is equivalent to letting the period τ become infinite. The index n is dropped because ω is now a continuous variable; the discrete sum becomes an integral and the function $f(t)$ is

$$f(t) = \frac{1}{2\pi} \int_{\omega=-\infty}^{\infty} F(\omega)d\omega \tag{7}$$

while the function $F(\omega)$ from (5) becomes

$$F(\omega) = \int_{u=-\infty}^{\infty} f(u)\exp(i\omega(t-u))du \tag{8}$$

On inserting $F(\omega)$ into (7) we get

$$\begin{aligned} f(t) &= \frac{1}{2\pi} \int_{\omega=-\infty}^{\infty} \left[\int_{u=-\infty}^{\infty} f(u)\exp(i\omega(t-u))du \right] d\omega \\ &= \frac{1}{2\pi} \int_{\omega=-\infty}^{\infty} \left[\int_{u=-\infty}^{\infty} f(u)\exp(-i\omega u)du \right] \exp(i\omega t)d\omega \end{aligned} \tag{9}$$

The quantity in square brackets, on changing the variable from u back to t, is

$$g(\omega) = \frac{1}{2\pi} \int_{t=-\infty}^{\infty} f(t)\exp(-i\omega t)dt \tag{10}$$

and the original expression can now be written

$$f(t) = \int_{\omega=-\infty}^{\infty} g(\omega)\exp(i\omega t)d\omega \tag{11}$$

The equivalence of these two equations is known as the Fourier integral theorem.

Fourier series that represent odd or even functions consist of sums of sines or cosines, respectively. In the same way, there are sine and cosine Fourier integrals that represent odd and even functions, respectively. Suppose that the function $f(t)$ is *even*, and let us replace the complex exponential in (1.272) using (1.5):

$$g(\omega) = \frac{1}{2\pi} \int_{t=-\infty}^{\infty} f(t)[\cos(\omega t) - i\sin(\omega t)]dt \qquad (1.273)$$

The sine function is odd, so, if $f(t)$ is even, the product $f(t)\sin(\omega t)$ is odd, and the integral of the second term is zero. The product $f(t)\cos(\omega t)$ is even, and we can convert the limits of integration to the positive interval:

$$g(\omega) = \frac{1}{2\pi} \int_{t=-\infty}^{\infty} f(t)\cos(\omega t)dt$$

$$= \frac{1}{\pi} \int_{t=0}^{\infty} f(t)\cos(\omega t)dt \qquad (1.274)$$

Thus, if $f(t)$ is even, then $g(\omega)$ is also even. Similarly, one finds that, if $f(t)$ is odd, $g(\omega)$ is also odd.

Now we expand the exponential in (1.271) and apply the same conditions of evenness and oddness to the products:

$$f(t) = \int_{\omega=-\infty}^{\infty} g(\omega)(\cos(\omega t) + i\sin(\omega t))d\omega$$

$$= 2 \int_{\omega=0}^{\infty} g(\omega)\cos(\omega t)d\omega \qquad (1.275)$$

If we were to substitute (1.275) back into (1.274), the integration would be preceded by a constant $2/\pi$, the product of the two constants in these equations. Equations (1.275) and (1.274) form a Fourier-transform pair, and it does not matter how the factor $2/\pi$ is divided between them. We will associate it here entirely with the second equation, so that we have the pair of equations

$$f(t) = \int_{\omega=0}^{\infty} g(\omega)\cos(\omega t)d\omega$$

$$g(\omega) = \frac{2}{\pi} \int_{t=0}^{\infty} f(t)\cos(\omega t)dt \qquad (1.276)$$

The even functions $f(t)$ and $g(\omega)$ are *Fourier cosine transforms* of each other.

A similar treatment for a function $f(t)$ that is *odd* leads to a similar pair of equations in which the Fourier transform $g(\omega)$ is also *odd* and

$$f(t) = \int_{\omega=0}^{\infty} g(\omega)\sin(\omega t)d\omega$$

$$g(\omega) = \frac{2}{\pi}\int_{t=0}^{\infty} f(t)\sin(\omega t)dt$$

(1.277)

The odd functions $f(t)$ and $g(\omega)$ are *Fourier sine transforms* of each other.

FURTHER READING

Boas, M. L. (2006). *Mathematical Methods in the Physical Sciences*, 3rd edn. Hoboken, NJ: Wiley, 839 pp.

James, J. F. (2004). *A Student's Guide to Fourier transforms*, 2nd edn. Cambridge: Cambridge University Press, 135 pp.

2

Gravitation

2.1 Gravitational acceleration and potential

The Universal Law of Gravitation deduced by Isaac Newton in 1687 describes the force of gravitational attraction between two point masses m and M separated by a distance r. Let a spherical coordinate system (r, θ, ϕ) be centered on the point mass M. The force of attraction $\mathbf{F}$ exerted on the point mass m acts radially inwards towards M, and can be written

$$\mathbf{F} = -G\frac{mM}{r^2}\mathbf{e}_r \tag{2.1}$$

In this expression, G is the gravitational constant ($6.674\,21 \times 10^{-11}$ m^3 kg^{-1} s^{-2}), $\mathbf{e}_r$ is the unit radial vector in the direction of increasing r, and the negative sign indicates that the force acts inwardly, towards the attracting mass. The gravitational acceleration $\mathbf{a}_G$ at distance r is the force on a unit mass at that point:

$$\mathbf{a}_G = -G\frac{M}{r^2}\mathbf{e}_r \tag{2.2}$$

The acceleration $\mathbf{a}_G$ may also be written as the negative gradient of a gravitational potential U_G

$$\mathbf{a}_G = -\nabla U_G \tag{2.3}$$

The gravitational acceleration for a point mass is radial, thus the potential gradient is given by

$$-\frac{\partial U_G}{\partial r} = -G\frac{M}{r^2} \tag{2.4}$$

$$U_G = -G\frac{M}{r} \tag{2.5}$$

In Newton's time the gravitational constant could not be verified in a laboratory experiment. The attraction between heavy masses of suitable dimensions is weak and the effects of friction and air resistance relatively large, so the first successful measurement of the gravitational constant by Lord Cavendish was not made until more than a century later, in 1798. However, Newton was able to confirm the validity of the inverse-square law of gravitation in 1687 by using existing astronomic observations of the motions of the planets in the solar system. These had been summarized in three important laws by Johannes Kepler in 1609 and 1619. The small sizes of the planets and the Sun, compared with the immense distances between them, enabled Newton to consider these as point masses and this allowed him to verify the inverse-square law of gravitation.

2.2 Kepler's laws of planetary motion

Johannes Kepler (1571–1630), a German mathematician and scientist, formulated his laws on the basis of detailed observations of planetary positions by Tycho Brahe (1546–1601), a Danish astronomer. The observations were made in the late sixteenth century, without the aid of a telescope. Kepler found that the observations were consistent with the following three laws (Fig. 2.1).

1. The orbit of each planet is an ellipse with the Sun at one focus.
2. The radius from the Sun to a planet sweeps over equal areas in equal intervals of time.

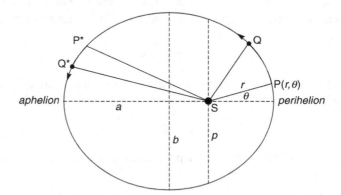

Fig. 2.1. Illustration of Kepler's laws of planetary motion. The orbit of each planet is an ellipse with the Sun at its focus (S); a, b, and p are the semi-major axis, semi-minor axis, and semi-latus rectum, respectively. The area swept by the radius to a planet in a given time is constant (i.e., area SPQ equals area SP*Q*); the square of the period is proportional to the cube of the semi-major axis. After Lowrie (2007).

3. The square of the period is proportional to the cube of the semi-major axis of the orbit.

The fundamental assumption is that the planets move under the influence of a central, i.e., radially directed force. For a planet of mass m at distance r from the Sun the force $\mathbf{F}$ can be written

$$\mathbf{F} = m\frac{d^2\mathbf{r}}{dt^2} = f(r)\mathbf{e}_r \tag{2.6}$$

The angular momentum $\mathbf{h}$ of the planet about the Sun is

$$\mathbf{h} = \mathbf{r} \times m\frac{d\mathbf{r}}{dt} \tag{2.7}$$

Differentiating with respect to time, the rate of change of angular momentum is

$$\frac{d\mathbf{h}}{dt} = m\frac{d}{dt}\left(\mathbf{r} \times \frac{d\mathbf{r}}{dt}\right) = m\left(\frac{d\mathbf{r}}{dt} \times \frac{d\mathbf{r}}{dt}\right) + m\left(\mathbf{r} \times \frac{d^2\mathbf{r}}{dt^2}\right) \tag{2.8}$$

The first term on the right-hand side is zero, because the vector product of a vector with itself (or with a vector parallel to itself) is zero. Thus

$$\frac{d\mathbf{h}}{dt} = \mathbf{r} \times m\frac{d^2\mathbf{r}}{dt^2} \tag{2.9}$$

On substituting from (2.6) and applying the same condition, we have

$$\frac{d\mathbf{h}}{dt} = \mathbf{r} \times f(r)\mathbf{e}_r = f(r)(\mathbf{r} \times \mathbf{e}_r) = 0 \tag{2.10}$$

This equation means that $\mathbf{h}$ is a constant vector; the angular momentum of the system is conserved. On taking the scalar product of $\mathbf{h}$ and $\mathbf{r}$, we obtain

$$\mathbf{r} \cdot \mathbf{h} = \mathbf{r} \cdot \left(\mathbf{r} \times m\frac{d\mathbf{r}}{dt}\right) \tag{2.11}$$

Rotating the sequence of the vectors in the triple product gives

$$\mathbf{r} \cdot \mathbf{h} = m\frac{d\mathbf{r}}{dt} \cdot (\mathbf{r} \times \mathbf{r}) = 0 \tag{2.12}$$

This result establishes that the vector $\mathbf{r}$ describing the position of a planet is always perpendicular to its constant angular momentum vector $\mathbf{h}$ and therefore defines a plane. Every planetary orbit is therefore a plane that passes through the Sun. The orbit of the Earth defines the *ecliptic* plane.

2.2.1 Kepler's Second Law

Let the position of a planet in its orbit be described by polar coordinates (r, θ) with respect to the Sun. The coordinates are defined so that the angle θ is zero at the closest approach of the planet to the Sun (perihelion). The angular momentum at an arbitrary point of the orbit has magnitude

$$h = mr^2 \frac{d\theta}{dt} \tag{2.13}$$

In a short interval of time Δt the radius vector from the Sun to the planet moves through a small angle $\Delta \theta$ and defines a small triangle. The area ΔA of the triangle is

$$\Delta A = \frac{1}{2} r^2 \Delta \theta \tag{2.14}$$

The rate of change of the area swept over by the radius vector is

$$\frac{dA}{dt} = \lim_{\Delta t \to 0} \left(\frac{\Delta A}{\Delta t} \right) = \lim_{\Delta t \to 0} \left(\frac{1}{2} r^2 \frac{\Delta \theta}{\Delta t} \right) \tag{2.15}$$

$$\frac{dA}{dt} = \frac{1}{2} r^2 \frac{d\theta}{dt} \tag{2.16}$$

On inserting from (2.13) we get

$$\frac{dA}{dt} = \frac{h}{2m} \tag{2.17}$$

Thus the area swept over by the radius vector in a given time is constant. This is Kepler's Second Law of planetary motion.

2.2.2 Kepler's First Law

If just the gravitational attraction of the Sun acts on the planet (i.e., we ignore the interactions between the planets), the total energy E of the planet is constant. The total energy E is composed of the planet's orbital kinetic energy and its potential energy in the Sun's gravitational field:

$$\frac{1}{2} m \left(\frac{dr}{dt} \right)^2 + \frac{1}{2} mr^2 \left(\frac{d\theta}{dt} \right)^2 - Gm \frac{S}{r} = E \tag{2.18}$$

The first term here is the planet's linear (radial) kinetic energy, the second term is its rotational kinetic energy (with mr^2 being the planet's moment of inertia about the Sun), and the third term is the gravitational potential energy. On writing

$$\frac{dr}{dt} = \frac{dr}{d\theta}\frac{d\theta}{dt} \qquad (2.19)$$

and rearranging terms we get

$$\left(\frac{dr}{d\theta}\right)^2\left(\frac{d\theta}{dt}\right)^2 + r^2\left(\frac{d\theta}{dt}\right)^2 - 2G\frac{S}{r} = 2\frac{E}{m} \qquad (2.20)$$

Now, to simplify later steps, we make a change of variables, writing

$$u = \frac{1}{r} \qquad (2.21)$$

Then

$$\frac{dr}{d\theta} = \frac{d}{d\theta}\left(\frac{1}{u}\right) = -\frac{1}{u^2}\left(\frac{du}{d\theta}\right) = -r^2\left(\frac{du}{d\theta}\right) \qquad (2.22)$$

Substituting from (2.22) into (2.20) gives

$$\left(r^2\frac{d\theta}{dt}\right)^2\left(\frac{du}{d\theta}\right)^2 + r^2\left(\frac{d\theta}{dt}\right)^2 - 2G\frac{S}{r} = 2\frac{E}{m} \qquad (2.23)$$

With the result of (2.13) we have

$$r^2\frac{d\theta}{dt} = \frac{h}{m}$$

$$r\left(\frac{d\theta}{dt}\right) = \frac{1}{r}\left(\frac{h}{m}\right) = u\left(\frac{h}{m}\right) \qquad (2.24)$$

On replacing these expressions, (2.23) becomes

$$\left(\frac{h}{m}\right)^2\left(\frac{du}{d\theta}\right)^2 + u^2\left(\frac{h}{m}\right)^2 - 2uGS = 2\frac{E}{m} \qquad (2.25)$$

$$\left(\frac{du}{d\theta}\right)^2 + u^2 - 2uGS\frac{m^2}{h^2} = 2\frac{Em}{h^2} \qquad (2.26)$$

The rest of the evaluation is straightforward, if painstaking. First we add a constant to each side,

$$\left(\frac{du}{d\theta}\right)^2 + u^2 - 2uGS\frac{m^2}{h^2} + \left(GS\frac{m^2}{h^2}\right)^2 = 2\frac{Em}{h^2} + \left(GS\frac{m^2}{h^2}\right)^2 \qquad (2.27)$$

$$\left(\frac{du}{d\theta}\right)^2 + \left(u - GS\frac{m^2}{h^2}\right)^2 = 2\frac{Em}{h^2} + \left(GS\frac{m^2}{h^2}\right)^2 \qquad (2.28)$$

Next, we move the second term to the right-hand side of the equation, giving

$$\left(\frac{du}{d\theta}\right)^2 = 2\frac{Em}{h^2} + \left(GS\frac{m^2}{h^2}\right)^2 - \left(u - GS\frac{m^2}{h^2}\right)^2 \qquad (2.29)$$

$$\left(\frac{du}{d\theta}\right)^2 = \left(GS\frac{m^2}{h^2}\right)^2 \left(1 + \frac{2Eh^2}{G^2S^2m^3}\right) - \left(u - GS\frac{m^2}{h^2}\right)^2 \qquad (2.30)$$

Now, we define some combinations of these terms, as follows:

$$u_0 = GS\frac{m^2}{h^2} \qquad (2.31)$$

$$e^2 = 1 + \frac{2Eh^2}{G^2S^2m^3} \qquad (2.32)$$

Using these defined terms, (2.30) simplifies to a more manageable form:

$$\left(\frac{du}{d\theta}\right)^2 = u_0^2 e^2 - (u - u_0)^2 \qquad (2.33)$$

$$\frac{du}{d\theta} = -\sqrt{u_0^2 e^2 - (u - u_0)^2} \qquad (2.34)$$

The solution of this equation, which can be tested by substitution, is

$$u = u_0(1 + e\cos\theta) \qquad (2.35)$$

The angle θ is defined to be zero at perihelion. The negative square root in (2.34) is chosen because, as θ increases, r increases and u must decrease. Let

$$p = \frac{1}{u_0} = \frac{h^2}{GSm^2} \qquad (2.36)$$

$$r = \frac{p}{1 + e\cos\theta} \qquad (2.37)$$

This is the polar equation of an ellipse referred to its focus, and is the proof of Kepler's First Law of planetary motion. The quantity e is the eccentricity of the ellipse, while p is the *semi-latus rectum* of the ellipse, which is half the length of a chord passing through the focus and parallel to the minor axis (Fig. 2.1).

These equations show that three types of trajectory around the Sun are possible, depending on the value of the total energy E in (2.18). If the kinetic energy is greater than the potential energy, the value of E in (2.32) is positive, and e is greater than 1; the path of the object is a hyperbola. If the kinetic energy and potential energy are equal, the total energy is zero and e is exactly 1; the path is a parabola. In each of these two cases the object can escape to infinity, and the paths are called escape trajectories. If the kinetic energy is less than the potential energy, the total energy E is negative and the eccentricity is less than 1. In this case (corresponding to a planet or asteroid) the object follows an elliptical orbit around the Sun.

2.2.3 Kepler's Third Law

It is convenient to describe the elliptical orbit in Cartesian coordinates (x, y), centered on the mid-point of the ellipse, instead of on the Sun. Define the x-axis parallel to the semi-major axis a of the ellipse and the y-axis parallel to the semi-minor axis b. The equation of the ellipse in Fig. 2.1 is

$$\frac{x^2}{a^2} + \frac{y^2}{b^2} = 1 \tag{2.38}$$

The semi-minor axis is related to the semi-major axis by the eccentricity e, so that

$$b^2 = a^2(1 - e^2) \tag{2.39}$$

The distance of the focus of the ellipse from its center is by definition ae. The length p of the semi-latus rectum is the value of y for a chord through the focus. On setting $y = p$ and $x = ae$ in (2.38), we obtain

$$\frac{p^2}{b^2} = 1 - \frac{(ae)^2}{a^2} = 1 - e^2 \tag{2.40}$$

$$p^2 = a^2(1 - e^2)^2 \tag{2.41}$$

Now consider the application of Kepler's Second Law to an entire circuit of the elliptical orbit. The area of the ellipse is πab, and the period of the orbit is T, so

$$\frac{dA}{dt} = \frac{\pi ab}{T} \tag{2.42}$$

Using (2.17),

$$\frac{h}{m} = \frac{2\pi ab}{T} \tag{2.43}$$

From (2.36) and (2.43) we get the value of the semi-latus rectum,

$$p = \frac{1}{GS}\left(\frac{h}{m}\right)^2 = \frac{1}{GS}\left(\frac{2\pi ab}{T}\right)^2 \tag{2.44}$$

Substituting from (2.41) gives

$$a(1 - e^2) = \frac{4\pi^2 a^2 b^2}{GST^2} = \frac{4\pi^2 a^4}{GST^2}(1 - e^2) \tag{2.45}$$

After simplifying, we finally get

$$\frac{T^2}{a^3} = \frac{4\pi^2}{GS} \tag{2.46}$$

The quantities on the right-hand side are constant, so the square of the period is proportional to the cube of the semi-major axis, which is Kepler's Third Law.

2.3 Gravitational acceleration and the potential of a solid sphere

The gravitational potential and acceleration outside and inside a solid sphere may be calculated from the Poisson and Laplace equations, respectively.

2.3.1 Outside a solid sphere, using Laplace's equation

Outside a solid sphere the gravitational potential U_G satisfies Laplace's equation (Section 1.9). If the density is uniform, the potential does not vary with the polar angle θ or azimuth ϕ. Under these conditions, Laplace's equation in spherical polar coordinates (2.67) reduces to

$$\frac{\partial}{\partial r}\left(r^2 \frac{\partial U_G}{\partial r}\right) = 0 \tag{2.47}$$

This implies that the bracketed quantity that we are differentiating must be a constant, C,

$$r^2 \frac{\partial U_G}{\partial r} = C \tag{2.48}$$

$$\frac{\partial U_G}{\partial r} = \frac{C}{r^2} \tag{2.49}$$

The gravitational acceleration outside the sphere is therefore

$$\mathbf{a}_G(r > R) = -\frac{\partial U_G}{\partial r} = -\left(\frac{C}{r^2}\right)\mathbf{e}_r \tag{2.50}$$

At its surface the gravitational acceleration has the value

$$\mathbf{a}_G(R) = -\frac{\partial U_G}{\partial r} = -\left(\frac{C}{R^2}\right)\mathbf{e}_r \tag{2.51}$$

The boundary condition at the surface of the sphere is that the accelerations determined outside and inside the sphere must be equal there. We use this to derive the value of the constant C. On comparing (2.51) and (2.60) we have

$$C = GM \tag{2.52}$$

On inserting for C in (2.50), the gravitational acceleration outside the sphere is

$$\mathbf{a}_G(r > R) = -G\frac{M}{r^2}\mathbf{e}_r \tag{2.53}$$

The gravitational potential outside the solid sphere is obtained by integrating (2.53) with respect to the radius. This gives

$$U_G(r > R) = -G\frac{M}{r} \tag{2.54}$$

2.3.2 Inside a solid sphere, using Poisson's equation

Inside a solid sphere with radius R and uniform density ρ the gravitational potential U_G satisfies Poisson's equation (Section 1.8). Symmetry again requires the use of spherical polar coordinates, and, because the density is uniform, there is no variation of potential with the polar angle θ or azimuth ϕ. Poisson's equation in spherical polar coordinates reduces to

$$\frac{1}{r^2}\frac{\partial}{\partial r}r^2\frac{\partial U_G}{\partial r} = 4\pi G\rho \tag{2.55}$$

On multiplying by r^2 and integrating with respect to r, we get

$$\frac{\partial}{\partial r}r^2\frac{\partial U_G}{\partial r} = 4\pi G\rho r^2 \tag{2.56}$$

$$r^2\frac{\partial U_G}{\partial r} = \frac{4}{3}\pi G\rho r^3 + C_1 \tag{2.57}$$

This equation has to be valid at the center of the sphere where $r = 0$, so the constant $C_1 = 0$ and

$$\frac{\partial U_G}{\partial r} = \frac{4}{3}\pi G\rho r \tag{2.58}$$

$$\mathbf{a}_G(r<R) = -\frac{\partial U_G}{\partial r} = \left(-\frac{4}{3}\pi G\rho r\right)\mathbf{e}_r \tag{2.59}$$

This shows that the gravitational acceleration inside a homogeneous solid sphere is proportional to the distance from its center. At the surface of the sphere, $r = R$, and the gravitational acceleration is

$$\mathbf{a}_G(R) = \left(-\frac{4}{3}\pi G\rho R\right)\mathbf{e}_r = -\frac{GM}{R^2}\mathbf{e}_r \tag{2.60}$$

where the mass M of the sphere is

$$M = \frac{4}{3}\pi R^3\rho \tag{2.61}$$

To obtain the *potential* inside the solid sphere, we must integrate (2.58). This gives

$$U_G = \frac{2}{3}\pi G\rho r^2 + C_2 \tag{2.62}$$

The constant of integration C_2 is obtained by noting that the potential must be continuous at the surface of the sphere. Otherwise a discontinuity would exist and the potential gradient (and force) would be infinite. Equating (2.54) and (2.62) at $r = R$ gives

$$\frac{2}{3}\pi G\rho R^2 + C_2 = -\frac{GM}{R} = -\frac{4}{3}\pi G\rho R^2 \tag{2.63}$$

$$C_2 = -2\pi G\rho R^2 \tag{2.64}$$

The gravitational potential inside the uniform solid sphere is therefore given by

$$U_G = \frac{2}{3}\pi G\rho r^2 - 2\pi G\rho R^2 \tag{2.65}$$

$$U_G = \frac{2}{3}\pi G\rho(r^2 - 3R^2) \tag{2.66}$$

A schematic graph of the variation of the gravitational potential inside and outside a solid sphere is shown in Fig. 2.2.

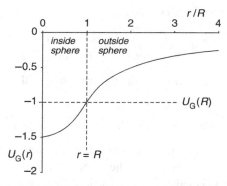

Fig. 2.2. Variation with radial distance r of the gravitational potential inside and outside a solid sphere of radius R. The potential of the surface of the sphere is $U_G(R)$.

2.4 Laplace's equation in spherical polar coordinates

In the above examples the sphere was assumed to have uniform density so that only the radial term in Laplace's equation had to be solved. This is also the case when density varies only with radius. In the Earth, however, lateral variations of the density distribution occur, and the gravitational potential U_G is then a solution of the full Laplace equation

$$\frac{1}{r^2}\frac{\partial}{\partial r}r^2\frac{\partial U_G}{\partial r} + \frac{1}{r^2\sin\theta}\frac{\partial}{\partial\theta}\sin\theta\frac{\partial U_G}{\partial\theta} + \frac{1}{r^2\sin^2\theta}\frac{\partial^2 U_G}{\partial\phi^2} = 0 \qquad (2.67)$$

This equation is solved using the method of *separation of variables*. This is a valuable mathematical technique, which allows the variables in a partial differential equation to be separated so that only terms in one variable are on one side of the equation and terms in other variables are on the opposite side. A trial solution for U_G is

$$U_G(r,\theta,\phi) = \Re(r) \cdot \Theta(\theta) \cdot \Phi(\phi) \qquad (2.68)$$

Here $\Re$, Θ, and Φ are all functions of a single variable only, namely r, θ, and ϕ, respectively. Multiplying (2.67) by r^2 and inserting (2.68) for U_G gives

$$\Theta\Phi\frac{\partial}{\partial r}r^2\frac{\partial\Re}{\partial r} + \frac{\Re\Phi}{\sin\theta}\frac{\partial}{\partial\theta}\sin\theta\frac{\partial\Theta}{\partial\theta} + \frac{\Re\Theta}{\sin^2\theta}\frac{\partial^2\Phi}{\partial\phi^2} = 0 \qquad (2.69)$$

On dividing throughout by $\mathfrak{R}\Theta\Phi$ we get

$$\frac{1}{\mathfrak{R}}\frac{\partial}{\partial r}r^2\frac{\partial\mathfrak{R}}{\partial r} + \frac{1}{\Theta\sin\theta}\frac{\partial}{\partial\theta}\sin\theta\frac{\partial\Theta}{\partial\theta} + \frac{1}{\Phi\sin^2\theta}\frac{\partial^2\Phi}{\partial\phi^2} = 0 \qquad (2.70)$$

Next we isolate the radial terms on the left-hand side of the equation, so that

$$\frac{1}{\mathfrak{R}}\frac{\partial}{\partial r}r^2\frac{\partial\mathfrak{R}}{\partial r} = -\frac{1}{\Theta\sin\theta}\frac{\partial}{\partial\theta}\sin\theta\frac{\partial\Theta}{\partial\theta} - \frac{1}{\Phi\sin^2\theta}\frac{\partial^2\Phi}{\partial\phi^2} \qquad (2.71)$$

The left-hand side of the equation is a function of r only, while the right-hand side does not depend on r. Whatever the value of the left-hand side, the right-hand side must always equal it. But r, θ, and ϕ are independent variables, so the identity can exist only if the opposite sides of the equation are equal to the same constant. Let this constant be K. For the opposite sides of (2.71) we get

$$\frac{1}{\mathfrak{R}}\frac{\partial}{\partial r}r^2\frac{\partial\mathfrak{R}}{\partial r} = K \qquad (2.72)$$

$$-\frac{1}{\Theta\sin\theta}\frac{\partial}{\partial\theta}\sin\theta\frac{\partial\Theta}{\partial\theta} - \frac{1}{\Phi\sin^2\theta}\frac{\partial^2\Phi}{\partial\phi^2} = K \qquad (2.73)$$

If we multiply the last equation throughout by $\sin^2\theta$, the variables can again be separated:

$$\frac{\sin\theta}{\Theta}\frac{\partial}{\partial\theta}\sin\theta\frac{\partial\Theta}{\partial\theta} + K\sin^2\theta = -\frac{1}{\Phi}\frac{\partial^2\Phi}{\partial\phi^2} \qquad (2.74)$$

The variables on the opposite sides of (2.74) are independent, so each side must be equal to the same constant, which we write temporarily as K_2. Thus we can replace equation (2.70) with three equations, consisting of (2.72) and the following two:

$$\frac{\sin\theta}{\Theta}\frac{\partial}{\partial\theta}\sin\theta\frac{\partial\Theta}{\partial\theta} + K\sin^2\theta = K_2 \qquad (2.75)$$

$$-\frac{1}{\Phi}\frac{\partial^2\Phi}{\partial\phi^2} = K_2 \qquad (2.76)$$

2.4.1 Azimuthal (longitudinal) solution

The constant K_2 may be chosen to suit the conditions governing the gravitational potential. The function $\Phi(\phi)$ describes the variation of the potential with azimuth (longitude, in geographic terms). If we measure azimuthal fluctuations of the potential around a circle of constant polar angle (geographic co-latitude), the same potential must result after a full circuit. This requires that the solution for $\Phi(\phi)$ be periodic, and that condition will be fulfilled if we let the constant equal m^2. For the right-hand side of (2.74) we get

$$-\frac{1}{\Phi}\frac{\partial^2\Phi}{\partial\phi^2} = m^2 \tag{2.77}$$

$$\frac{\partial^2\Phi}{\partial\phi^2} + m^2\Phi = 0 \tag{2.78}$$

This is the equation of simple harmonic motion, which has periodic solutions of the form

$$\Phi(\phi) = a_m\cos(m\phi) + b_m\sin(m\phi) \tag{2.79}$$

2.4.2 Polar (latitudinal) solution for rotational symmetry

We first consider solutions of Laplace's equation that have rotational symmetry about the reference axis, which in the Earth is its axis of rotation. Since there is no azimuthal variation of the potential in this situation, we can set $m = 0$. The variation of potential with angle θ is described by

$$\sin\theta\,\frac{\partial}{\partial\theta}\sin\theta\,\frac{\partial\Theta}{\partial\theta} + \left(K\sin^2\theta\right)\Theta = 0 \tag{2.80}$$

$$\frac{1}{\sin\theta}\frac{\partial}{\partial\theta}\sin\theta\,\frac{\partial\Theta}{\partial\theta} + K\Theta = 0 \tag{2.81}$$

$$\left(\frac{-1}{\sin\theta}\frac{\partial}{\partial\theta}\right)\sin^2\theta\left(\frac{-1}{\sin\theta}\frac{\partial\Theta}{\partial\theta}\right) + K\Theta = 0 \tag{2.82}$$

If we write $x = \cos\theta$, then

$$\frac{\partial}{\partial x} = \frac{-1}{\sin\theta}\frac{\partial}{\partial\theta} \tag{2.83}$$

and (2.82) becomes

$$\frac{\partial}{\partial x}\left[(1 - x^2)\frac{\partial\Theta}{\partial x}\right] + K\Theta = 0 \tag{2.84}$$

Comparison with (1.175) shows that this is equivalent to the Legendre differential equation, with $n(n + 1) = K$. If we make this choice of constant, we ensure that the Laplace equation will have periodic solutions in polar angle (co-latitude), namely the Legendre polynomials. The equation is

$$\frac{\partial}{\partial x}\left[(1-x^2)\frac{\partial P_n(x)}{\partial x}\right] + n(n+1)P_n(x) = 0 \tag{2.85}$$

and its solutions are

$$\Theta_n = P_n(x) = P_n(\cos\theta) \tag{2.86}$$

2.4.3 Radial solution

With $K = n(n+1)$, the equation for the radial variation of the gravitational potential becomes

$$\frac{1}{\Re}\frac{\partial}{\partial r}r^2\frac{\partial \Re}{\partial r} = n(n+1) \tag{2.87}$$

There will be a radial solution for each value of n, so we write it $\Re_n$, where

$$\frac{\partial}{\partial r}r^2\frac{\partial \Re_n}{\partial r} - n(n+1)\Re_n = 0 \tag{2.88}$$

Let $\Re_n(r)$ be represented by the power series

$$\Re_n(r) = \sum_{p=0}^{\infty} a_p r^p \tag{2.89}$$

Differentiating with respect to r gives

$$\frac{\partial \Re}{\partial r} = \sum_{p=0}^{\infty} p a_p r^{p-1} \tag{2.90}$$

Multiplying by r^2 and differentiating the product

$$r^2\frac{\partial \Re}{\partial r} = \sum_{p=0}^{\infty} p a_p r^{p+1} \tag{2.91}$$

$$\frac{\partial}{\partial r}r^2\frac{\partial \Re}{\partial r} = \sum_{p=0}^{\infty} p(p+1) a_p r^p \tag{2.92}$$

Inserting this result into (2.88) gives

$$\sum_{p=0}^{\infty} p(p+1) a_p r^p - n(n+1) \sum_{p=0}^{\infty} a_p r^p = 0 \qquad (2.93)$$

$$\sum_{p=0}^{\infty} a_p r^p [p(p+1) - n(n+1)] = 0 \qquad (2.94)$$

For this result to be true for any value of r, the expression in square brackets must equal zero,

$$p(p+1) - n(n+1) = 0 \qquad (2.95)$$

That is,

$$p^2 + p - n(n+1) = 0 \qquad (2.96)$$

Thus p can have the values $p = n$ or $p = -(n+1)$ and the radial variation of the potential is given by

$$\Re_n(r) = A_n r^n + \frac{B_n}{r^{n+1}} \qquad (2.97)$$

where A_n and B_n are constants determined by the boundary conditions.

2.4.4 Solution of Laplace's equation for rotational symmetry

Combining the radial and polar variations, the gravitational potential for a mass distribution that has rotational symmetry about an axis is

$$U_G = \sum_{n=0}^{\infty} \left(A_n r^n + \frac{B_n}{r^{n+1}} \right) P_n(\cos\theta) \qquad (2.98)$$

2.4.5 General solution of Laplace's equation

In the general case the potential may vary azimuthally about the reference axis. The constant m is no longer zero and instead of (2.80) we have

$$\frac{\sin\theta}{\Theta} \frac{\partial}{\partial\theta} \sin\theta \frac{\partial\Theta}{\partial\theta} + K \sin^2\theta = m^2 \qquad (2.99)$$

$$\sin\theta \frac{\partial}{\partial\theta} \sin\theta \frac{\partial\Theta}{\partial\theta} + (K \sin^2\theta - m^2)\Theta = 0 \qquad (2.100)$$

As in the case with rotational symmetry, we substitute $x = \cos\theta$ and obtain

$$\frac{\partial}{\partial x}\left(1 - x^2\right)\frac{\partial \Theta}{\partial x} + \left[K_1 - \frac{m^2}{1 - x^2}\right]\Theta = 0 \tag{2.101}$$

If we again write $n(n+1)$ for the constant K,

$$\left(1 - x^2\right)\frac{\partial^2 \Theta}{\partial x^2} - 2x\frac{\partial \Theta}{\partial x} + \left[n(n+1) - \frac{m^2}{1 - x^2}\right]\Theta = 0 \tag{2.102}$$

This equation is equivalent to the associated Legendre equation (1.237), and the functions Θ are the associated Legendre polynomials:

$$\Theta(\theta) = P_n^m(x) = P_n^m(\cos\theta) \tag{2.103}$$

The general solution of Laplace's equation for the gravitational potential in spherical polar coordinates is obtained by combining the results of (2.79), (2.97), and (2.103):

$$U_G = \sum_{n=0}^{\infty}\sum_{m=0}^{n}\left(A_n r^n + \frac{B_n}{r^{n+1}}\right)\left(a_n^m\cos(m\phi) + b_n^m\sin(m\phi)\right)P_n^m(\cos\theta) \tag{2.104}$$

2.5 MacCullagh's formula for the gravitational potential

The yielding of the Earth to the deforming forces of its own rotation results in a shape that is symmetric about the rotation axis and slightly flattened at the poles. The figure is classified as an ellipsoid of revolution, and, since it deviates only slightly from a sphere, it may be called a spheroid. The equation and geometric properties of a spheroid are summarized in Box 2.1.

The flattening of the Earth is defined as the difference between the equatorial radius and the polar radius, expressed as a fraction of the equatorial radius:

$$f = \frac{a - c}{a} \tag{2.105}$$

The value of f is known accurately from satellite geodesy (Table 2.1) to be $f = 1/298.252$.

Let the Earth be represented by a spheroid with flattening f, and let the origin of a Cartesian coordinate system (x, y, z) be at the center of mass of the spheroid (Fig. 2.3). U_G is the gravitational potential at an external point P at distance r from the center of the Earth. For a continuous distribution of mass in a body we can employ integral calculus to calculate its mass, moments of inertia, or the location of its center of mass. However, it is instructive to regard the Earth as a

Box 2.1.   The ellipsoid and spheroid

Let an ellipsoid with three unequal principal axes be referred to a set of
orthogonal Cartesian axes (x, y, z) such that the x-axis is oriented parallel to
the longest dimension of the ellipsoid and the z-axis parallel to its shortest
dimension (Fig. B2.1(a)). The equation of the ellipsoid is

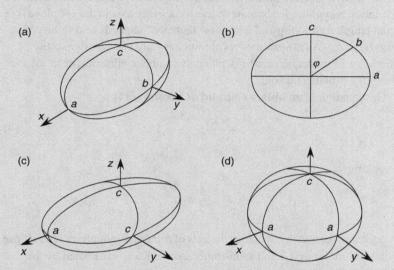

Fig. B2.1. (a) General ellipsoid with three unequal principal axes, $a > b > c$; (b)
elliptical cross-section through the center of an ellipsoid; b is the radius of a
circular section, inclined to the short axis c at an angle φ; (c) prolate ellipsoid;
and (d) oblate ellipsoid.

$$\frac{x^2}{a^2} + \frac{y^2}{b^2} + \frac{z^2}{c^2} = 1 \tag{1}$$

where a, b, and c – the intercepts of the ellipsoid with the x, y, and z reference
axes, respectively – are the lengths of its principal axes. The volume of the
ellipsoid is

$$V = \frac{4}{3}\pi abc \tag{2}$$

Each cross-section through the center of a triaxial ellipsoid is an ellipse,
except for two, which are circular sections. Defining the axes such that $a > b$
$> c$, the radius of a circular section is equal to the intermediate axis b and it is
inclined to the short axis c at an angle φ (Fig. B2.1(b)) given by

$$\tan\varphi = \frac{a}{c}\sqrt{\frac{b^2-c^2}{a^2-b^2}} \tag{3}$$

An **ellipsoid of revolution** is symmetric about one of its axes. If this is the long x-axis, every axis in the y–z plane is of equal length c. An ellipsoid with this elongated shape is said to be *prolate* (Fig. B2.1(c)). If the ellipsoid of revolution is symmetric about its short z-axis, every axis in the x–y plane is of equal length a. An ellipsoid with this "flattened" shape is said to be *oblate* (Fig. B2.1(d)). An ellipsoid of revolution has only one circular section, which lies in the (equatorial) x–y plane of an oblate ellipsoid, or in the y–z plane of a prolate ellipsoid.

The equation of an **oblate ellipsoid of revolution** is

$$\frac{x^2+y^2}{a^2}+\frac{z^2}{c^2}=1 \tag{4}$$

Its volume is

$$V=\frac{4}{3}\pi a^2 c \tag{5}$$

Every cross-section that includes the axis of rotational symmetry is an ellipse with semi-major axis a and semi-minor axis c. These are related by the *ellipticity, f*, defined as

$$f=\frac{a-c}{a} \tag{6}$$

An oblate ellipsoid of revolution that is almost spherical in shape (i.e., the axes a and c are almost equal) is called a **spheroid**. This is the closest geometric approximation to the shape of the Earth; the ellipticity of a polar section of the spheroid is called the *flattening*.

collection of discrete point masses m_i like the one at Q with Cartesian coordinates (x_i, y_i, z_i). This point mass is distant r_i from the center and u_i from the observation point at P. The gravitational potential at P can be written (compare with (2.54)) as the sum of contributions from all the point masses in the body:

$$U_G = -G\sum_i \frac{m_i}{u_i} \tag{2.106}$$

Table 2.1. *Some useful geodetic parameters (source: Groten, 2004)*

Parameter	Symbol	Units	Value
Geocentric gravitational constant	GE	$10^{14} \, \text{m}^3 \, \text{s}^{-2}$	3.986 004 418
Mass of Earth: GE/G	E	$10^{24} \, \text{kg}$	5.973 7
Equatorial radius	a	km	6,378.136 7
Polar radius: $a(1-f)$	c	km	6,356.752
Radius of equivalent sphere: $(a^2 c)^{1/3}$	R	km	6,371.000 4
Flattening	f	10^{-3}	3.352 865 9
Inverse flattening	$1/f$		298.252 31
Dynamic form factor	J_2	10^{-3}	1.082 635 9
Nominal mean angular velocity	Ω	$10^{-5} \, \text{rad s}^{-1}$	7.292 115
Mean equatorial gravity	g_e	m s^{-2}	9.780 327 8
Acceleration ratio: $\Omega^2 a^3/(GE)$	m	10^{-3}	3.461 391
Inverse acceleration ratio	$1/m$		288.901
Moment of inertia ratio for C	$C/(Ea^2)$		0.330 701
Moment of inertia ratio for B	$B/(Ea^2)$		0.329 622
Moment of inertia ratio for A	$A/(Ea^2)$		0.329 615
Dynamic ellipticity	H	10^{-3}	3.273 787 5
Inverse dynamic ellipticity	$1/H$		304.513

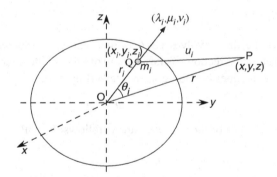

Fig. 2.3. Configuration for calculation of the gravitational potential of an ellipsoid, considered as a distribution of discrete point masses m_i.

Let the radius to the point mass at Q make an angle θ_i with the radius to the external point P. The reciprocal distance formula (1.157) for the Legendre polynomials can be applied to the sides of the triangle OPQ:

$$\frac{1}{u_i} = \frac{1}{r} \sum_{n=0}^{\infty} \left(\frac{r_i}{r}\right)^n P_n(\cos\theta_i) \tag{2.107}$$

Substituting this into (2.106) gives for the gravitational potential of the body

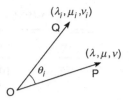

Fig. 2.4. Angle θ_i bounded by straight lines OP, with direction cosines (λ, μ, ν), and OQ, with direction cosines $(\lambda_i, \mu_i, \nu_i)$.

$$U_G = -G \sum_i m_i \frac{1}{r} \sum_{n=0}^{\infty} \left(\frac{r_i}{r}\right)^n P_n(\cos \theta_i) \tag{2.108}$$

Expanding the reciprocal distance formula gives an infinite sequence of terms. The ratio of successive terms depends on r_i/r, which is less than 1 outside the body. Moreover, if the shape of the body does not deviate much from a sphere, higher-order terms are not significant, so

$$U_G \approx -G\frac{1}{r}\sum_i m_i - G\frac{1}{r^2}\sum_i m_i r_i \cos\theta_i - G\frac{1}{r^3}\sum_i m_i r_i^2 P_2(\cos\theta_i)$$
$$= U_0 + U_1 + U_2 \tag{2.109}$$

Each term after the first involves $\cos\theta_i$, which can be computed (Box 1.2, equation (6)) from the direction cosines (λ, μ, ν) of OP and the direction cosines $(\lambda_i, \mu_i, \nu_i)$ of OQ, the lines bounding the angle θ_i (Fig. 2.4):

$$\cos\theta_i = \lambda\lambda_i + \mu\mu_i + \nu\nu_i \tag{2.110}$$

The direction cosines of the two lines are as follows: for OP,

$$\lambda = \frac{x}{r}, \qquad \mu = \frac{y}{r}, \qquad \nu = \frac{z}{r} \tag{2.111}$$

and for OQ,

$$\lambda_i = \frac{x_i}{r_i}, \qquad \mu_i = \frac{y_i}{r_i}, \qquad \nu_i = \frac{z_i}{r_i} \tag{2.112}$$

Substituting into (2.110) gives

$$\cos\theta_i = \frac{1}{rr_i}(xx_i + yy_i + zz_i) \tag{2.113}$$

Now we take a closer look at the individual terms in (2.109) for the potential. For the case $n = 0$, potential U_0:

$$U_0 = -G\frac{1}{r}\sum_i m_i = -\frac{GM}{r} \tag{2.114}$$

Comparison with (2.54) shows that U_0 is the potential of a sphere at an external point P.

For the case $n = 1$, potential U_1:

$$U_1 = -G\frac{1}{r^2}\sum_i m_i r_i \cos\theta_i \tag{2.115}$$

From (2.113) we obtain

$$r_i \cos\theta_i = \frac{1}{r}(x x_i + y y_i + z z_i) \tag{2.116}$$

On substituting into (2.115) and gathering terms, we have

$$U_1 = -G\frac{1}{r^3}\left[x\sum_i m_i x_i + y\sum_i m_i y_i + z\sum_i m_i z_i\right] \tag{2.117}$$

The origin of the coordinate system is at the center of mass of the body. The center of mass is defined as the point about which the sums of the moments of the point masses that make up the body are zero:

$$\sum_i m_i x_i = \sum_i m_i y_i = \sum_i m_i z_i = 0 \tag{2.118}$$

Each sum on the right-hand side of (2.117) is zero, and consequently

$$U_1 = 0 \tag{2.119}$$

For the case $n = 2$, potential U_2:

$$U_2 = -G\frac{1}{r^3}\sum_i m_i r_i^2 P_2(\cos\theta_i) \tag{2.120}$$

On substituting for $P_2(\cos\theta)$ from Table 1.1, we obtain

$$U_2 = -G\frac{1}{2r^3}\sum_i m_i r_i^2 (3\cos^2\theta_i - 1) = -G\frac{1}{2r^3}\sum_i m_i r_i^2 (2 - 3\sin^2\theta_i) \tag{2.121}$$

$$U_2 = -G\frac{1}{2r^3}\left[\sum_i 2m_i r_i^2 - 3\sum_i m_i r_i^2 \sin^2\theta_i\right] \tag{2.122}$$

The principal moments of inertia A, B, and C of a body about the x-, y-, and z-axes, respectively, are defined in Box 2.2:

Box 2.2.   **Moments and products of inertia**

The angular momentum h of a body rotating at angular velocity ω about an axis is given by

$$h = I\omega \tag{1}$$

The quantity I is the *moment of inertia* of the body. It is a measure of the distribution of its mass about an axis of rotation. For a point mass m at perpendicular distance r from an axis of rotation the moment of inertia is

$$I = mr^2 \tag{2}$$

If an extended body is made up of discrete particles with mass m_i at distance r_i from the rotation axis, the moment of inertia is the sum of all the contributions from all these particles:

$$I = \sum_i m_i r_i^2 \tag{3}$$

Let the mass distribution of a body be described relative to three orthogonal Cartesian coordinate axes. The moments of inertia A, B, and C about the x-, y-, and z-axes, respectively, are

$$A = \sum_i m_i (y_i^2 + z_i^2)$$
$$B = \sum_i m_i (z_i^2 + x_i^2) \tag{4}$$
$$C = \sum_i m_i (x_i^2 + y_i^2)$$

Another property that affects the rotational behavior of a body is its *product of inertia* about the axis of rotation. The products of inertia H, J, and K of a body relative to the x-, y-, and z- reference axes are defined as

$$H = \sum_i m_i y_i z_i$$
$$J = \sum_i m_i z_i x_i \tag{5}$$
$$K = \sum_i m_i x_i y_i$$

Suppose that in a homogeneous body the z–x plane is a plane of symmetry. For every particle at (x_i, y_i) there is an equivalent particle at $(x_i, -y_i)$ that cancels out its contribution to the product of inertia K, which is therefore zero. If each pair of reference axes defines a plane of symmetry – as in a sphere, spheroid, or ellipsoid – then all the products of inertia are zero. Non-zero products of inertia are expressions of the lack of symmetry of a homogeneous body.

$$A = \sum_i m_i\left(y_i^2 + z_i^2\right), \qquad B = \sum_i m_i\left(z_i^2 + x_i^2\right), \qquad C = \sum_i m_i\left(x_i^2 + y_i^2\right)$$

$$(2.123)$$

Adding these moments of inertia gives

$$A + B + C = 2\sum_i m_i r_i^2 \qquad (2.124)$$

Substituting into (2.122) gives

$$U_2 = -G\frac{1}{2r^3}\left[A + B + C - 3\sum_i m_i r_i^2 \sin^2\theta_i\right] \qquad (2.125)$$

Let the moment of inertia of the body about the line OP joining the center of the ellipsoid and the point of observation be I (Box 2.2). The distance of the point Q from the line OP (Fig. 2.3) is $r_i \sin\theta_i$ and the moment of inertia I is given by

$$I = \sum_i m_i r_i^2 \sin^2\theta_i \qquad (2.126)$$

The second-order term in the potential becomes

$$U_2 = -G\frac{1}{2r^3}(A + B + C - 3I) \qquad (2.127)$$

Combining the expressions for U_0 and U_2, the gravitational potential of the spheroid at P is

$$U_G = -G\frac{M}{r} - G\frac{A + B + C - 3I}{2r^3} \qquad (2.128)$$

This is known as MacCullagh's formula (and dates from 1855).

2.5.1 Gravitational potential of a spheroid

The shape of the Earth deviates only slightly from a sphere and is best represented as a spheroid that is symmetric about the rotation axis. For an ellipsoid the moment of inertia I in MacCullagh's formula can be expressed in terms of the principal moments of inertia A, B, and C. The definition of I can be expanded as

$$I = \sum_i m_i r_i^2 \sin^2\theta_i = \sum_i m_i r_i^2 - \sum_i m_i r_i^2 \cos^2\theta_i \qquad (2.129)$$

Because the sum of the squares of direction cosines is 1, we can write

$$\sum_i m_i r_i^2 = \sum_i m_i \left(x_i^2 + y_i^2 + z_i^2\right)\left(\lambda^2 + \mu^2 + \nu^2\right) \tag{2.130}$$

Using the definitions of $r_i \cos \theta_i$ (2.116) and the direction cosines (λ, μ, ν) of OP (2.111),

$$\sum_i m_i r_i^2 \cos^2 \theta_i = \frac{1}{r^2} \sum_i m_i (x x_i + y y_i + z z_i)^2$$
$$= \sum_i m_i (\lambda x_i + \mu y_i + \nu z_i)^2 \tag{2.131}$$

Expanding the squared expression and taking the direction cosines outside the sums gives

$$\sum_i m_i r_i^2 \cos^2 \theta_i = \lambda^2 \sum_i m_i x_i^2 + \mu^2 \sum_i m_i y_i^2 + \nu^2 \sum_i m_i z_i^2$$
$$+ 2\lambda\mu \sum_i m_i x_i y_i + 2\mu\nu \sum_i m_i y_i z_i + 2\nu\lambda \sum_i m_i z_i x_i \tag{2.132}$$

On combining (2.130) and (2.132), we have that the moment of inertia of the ellipsoid about the line OP is

$$I = \lambda^2 \sum_i m_i \left(y_i^2 + z_i^2\right) + \mu^2 \sum_i m_i \left(z_i^2 + x_i^2\right) + \nu^2 \sum_i m_i \left(x_i^2 + y_i^2\right)$$
$$- 2\lambda\mu \sum_i m_i x_i y_i - 2\mu\nu \sum_i m_i y_i z_i - 2\nu\lambda \sum_i m_i z_i x_i \tag{2.133}$$

The first three sums on the right are recognizable as the definitions of the principal moments of inertia A, B, and C, while the final three terms are definitions of the products of inertia H, J, and K (see Box 2.2). Thus the moment of inertia I about an axis with direction cosines (λ, μ, ν) is related to the principal moments and products of inertia by

$$I = A\lambda^2 + B\mu^2 + C\nu^2 - 2K\lambda\mu - 2H\mu\nu - 2J\nu\lambda \tag{2.134}$$

In an ellipsoid the x–y, y–z, and z–x planes are planes of symmetry, so the products of inertia are $H = J = K = 0$. The expression for I reduces in the case of an ellipsoid to

$$I = A\lambda^2 + B\mu^2 + C\nu^2 \tag{2.135}$$

Substituting this expression for I in MacCullagh's formula gives

$$U_G = -G\frac{M}{r} - G\left(\frac{A + B + C - 3\left(A\lambda^2 + B\mu^2 + C\nu^2\right)}{2r^3}\right) \tag{2.136}$$

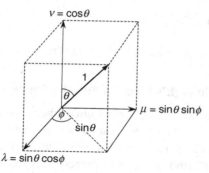

Fig. 2.5. Relationship between the direction cosines of a line and the angles θ and ϕ that define its direction.

The symmetry of the Earth about its rotation axis means that the moment of inertia about any axis in the equatorial plane has the same value, i.e., $A = B$. For the spheroidal Earth this results in

$$U_G = -G\frac{M}{r} - G\left(\frac{2A + C - 3A(\lambda^2 + \mu^2) - 3C\nu^2}{2r^3}\right) \qquad (2.137)$$

Now we revert from the direction cosines of OP to the direction of the line in terms of the angles θ and ϕ, corresponding respectively to co-latitude and longitude in geographic terms. These angles and the direction cosines are related as in Fig. 2.5:

$$\lambda = \sin\theta\cos\phi$$
$$\mu = \sin\theta\sin\phi \qquad (2.138)$$
$$\nu = \cos\theta$$

Squaring and summing the direction cosines λ and μ gives

$$\lambda^2 + \mu^2 = \sin^2\theta(\cos^2\phi + \sin^2\phi) = \sin^2\theta$$
$$= 1 - \cos^2\theta \qquad (2.139)$$

Replacing the direction cosines with the above expressions gives

$$U_G = -G\frac{M}{r} - G\left(\frac{2A + C - 3A(1 - \cos^2\theta) - 3C\cos^2\theta}{2r^3}\right) \qquad (2.140)$$

$$U_G = -G\frac{M}{r} - G(C - A)\left(\frac{1 - 3\cos^2\theta}{2r^3}\right) \qquad (2.141)$$

$$U_G = -G\frac{M}{r} + G\frac{C-A}{r^3}P_2(\cos\theta) \tag{2.142}$$

This is the gravitational potential of an ellipsoid of revolution at an external point.

2.5.2 MacCullagh's formula and the figure of the Earth

The Earth's shape deviates only slightly from a sphere, and is close to that of an oblate spheroid. MacCullagh's formula is not an exact expression for the gravitational potential of the Earth, because terms of higher order than U_2 were omitted from (2.109). In order to express U_G more exactly, we need to use an infinite series of potentials:

$$U_G = U_0 + U_1 + U_2 + U_3 + \cdots = \sum_{n=0}^{\infty} U_n \tag{2.143}$$

Each term of order n is proportional to $(1/r)^n$ and decreases in relative importance with increasing distance r. An alternative form for the gravitational potential U_G of the Earth at an external point is to write it as an infinite series of terms involving the Legendre polynomials and using Earth's mass E and equatorial radius a:

$$U_G = -G\frac{E}{r}\left[1 - \sum_{n=2}^{\infty} J_n\left(\frac{a}{r}\right)^n P_n(\cos\theta)\right] \tag{2.144}$$

The sum inside the square brackets modifies the potential U_0 of a sphere to reflect the real mass distribution in the Earth. The coefficients J_n describe the relative importance of successive terms in the series. The sum begins at $n = 2$ because $U_1 = 0$ when the coordinate system is centered at the Earth's center of mass, as in (2.119). Values for the coefficients J_n are obtained from satellite geodesy. They are very small, of order 10^{-6}, except for J_2, which is about 1,000 times larger and has the value 1.082×10^{-3}. J_2 is called the *dynamic form factor* of the Earth. The coefficient J_3 has the value -2.54×10^{-6}; it describes a slight deviation from a spheroid, being more depressed at the south pole and elevated at the north pole. This makes the Earth slightly pear-shaped. The coefficient J_4 is equal to -1.59×10^{-6} and is needed in order to obtain a more exact description of the gravitational potential for a model Earth whose mass distribution is symmetric about the equator.

Writing (2.144) to first order:

$$U_G = -G\frac{E}{r}\left[1 - J_2\left(\frac{a}{r}\right)^2 P_2(\cos\theta)\right] \tag{2.145}$$

This has to be equivalent to MacCullagh's formula for the spheroidal Earth. On equating terms in (2.142) and (2.145), we get the result

$$G\frac{E}{r}J_2\left(\frac{a}{r}\right)^2 P_2(\cos\theta) = G\frac{C-A}{r^3}P_2(\cos\theta) \qquad (2.146)$$

where

$$J_2 = \frac{C-A}{Ea^2} \qquad (2.147)$$

This result shows that the dynamic form factor J_2 is dependent on the difference between the principal moments of inertia, C and A. The polar flattening of Earth's figure results from the centrifugal acceleration of its rotation. The redistribution of mass finds expression as a difference between the principal moments of inertia. This difference, in turn, affects how the Earth reacts to external gravitational torques, which cause the rotation axis to precess about the pole to the ecliptic. The difference between C and A even affects the free rotation of the Earth, creating a longer-period wobble that is superposed on the daily rotation.

FURTHER READING

Blakely, R. J. (1995). *Potential Theory in Gravity & Magnetic Applications*. Cambridge: Cambridge University Press, 441 pp.

Lowrie, W. (2007). *Fundamentals of Geophysics*, 2nd edn. Cambridge: Cambridge University Press, 381 pp.

Officer, C. B. (1974). *Introduction to Theoretical Geophysics*. New York: Springer, 385 pp.

Stacey, F. D. and Davis, P. M. (2008). *Physics of the Earth*, 4th edn. Cambridge: Cambridge University Press, 532 pp.

3

Gravity

At any point on the Earth gravity acts in a direction normal to a surface on which the potential of gravity is constant. This equipotential surface is the best-fitting geometric figure to mean sea-level on the Earth. Its shape is that of a slightly flattened spheroid, for which the radius at any point can be computed. The potential of gravity on this spheroid – the *geopotential* – is computed by combining the gravitational potential and the potential of the centrifugal acceleration due to Earth's rotation. Gravity measurements are made with a high degree of accuracy. In order to compute a theoretical value of gravity for comparison at any latitude similar accuracy must be attained. Consequently, each step in computing the formula for the reference gravity must be carried out to second order in the flattening f and related parameters.

3.1 The ellipticity of the Earth's figure

Every cross-section of Earth's spheroidal shape that includes both poles is an identical ellipse, with equatorial semi-major axis a and polar semi-minor axis c, which are related (Box 2.1) by the flattening f through the equation $c = a(1 - f)$. In Cartesian coordinates the equation of the ellipse is

$$\frac{x^2}{a^2} + \frac{z^2}{c^2} = 1 \tag{3.1}$$

A position on the reference spheroid is specified by the polar angle θ and radius r, defined relative to the axis of rotational symmetry and center of the spheroid, repectively (Fig. 3.1). Consider a polar cross-section that includes the x- and z-axes, so that $x = r \sin \theta$ and $z = r \cos \theta$. By substituting into (3.1) we get the equation of the elliptical section in polar coordinates:

$$\frac{r^2 \sin^2\theta}{a^2} + \frac{r^2 \cos^2\theta}{c^2} = 1 \tag{3.2}$$

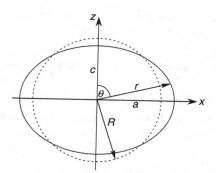

Fig. 3.1. Polar cross-section of a spheroid with principal axes a and c ($c<a$), compared with a sphere (dashed) with radius R and the same volume as the spheroid.

$$\frac{r^2}{a^2}\left(\sin^2\theta + \frac{\cos^2\theta}{(1-f)^2}\right) = 1 \tag{3.3}$$

On rearranging slightly, this becomes

$$r^2 = \frac{a^2(1-f)^2}{\cos^2\theta + (1-f)^2\sin^2\theta} \tag{3.4}$$

The denominator can be expanded, giving

$$\cos^2\theta + (1-f)^2\sin^2\theta = (1 - 2f + f^2)\sin^2\theta + \cos^2\theta$$
$$= \sin^2\theta + \cos^2\theta - 2f\sin^2\theta + f^2\sin^2\theta \tag{3.5}$$

Noting that $\sin^2\theta + \cos^2\theta = 1$, we can rewrite this as

$$\cos^2\theta + (1-f)^2\sin^2\theta = 1 - 2f\sin^2\theta + f^2\sin^2\theta(\sin^2\theta + \cos^2\theta)$$
$$= 1 - 2f\sin^2\theta + f^2\sin^4\theta + f^2\sin^2\theta\cos^2\theta \tag{3.6}$$
$$= (1 - f\sin^2\theta)^2 + f^2\sin^2\theta\cos^2\theta$$

By substituting into (3.4) and taking the square root, we get an equation for the radius:

$$\frac{r}{a} = \frac{1-f}{\left((1 - f\sin^2\theta)^2 + f^2\sin^2\theta\cos^2\theta\right)^{1/2}}$$
$$= \frac{1-f}{1 - f\sin^2\theta}\left(1 + \frac{f^2\sin^2\theta\cos^2\theta}{(1 - f\sin^2\theta)^2}\right)^{-1/2} \tag{3.7}$$

Applying the binomial theorem twice to the last line and expanding to order f^2 gives an equation for the surface of a spheroid

$$\frac{r}{a} \approx \frac{1-f}{1-f\sin^2\theta}\left(1 - \frac{1}{2}f^2\sin^2\theta\cos^2\theta\right) \approx \frac{1-f}{1-f\sin^2\theta} \tag{3.8}$$

The expansions for the gravitational potential and for gravity on the reference ellipsoid require the ratio a/r. Upon inverting (3.8) with the aid of the binomial expansion we get, to order f^2,

$$\frac{a}{r} \approx \frac{1-f\sin^2\theta}{1-f}\left(1 + \frac{1}{2}f^2\sin^2\theta\cos^2\theta\right)$$

$$\approx \left(1 - f\sin^2\theta\right)\left(1 + \frac{1}{2}f^2\sin^2\theta\cos^2\theta\right)\left(1 + f + f^2 + \cdots\right)$$

$$\approx 1 + f + f^2 - f\sin^2\theta - f^2\sin^2\theta + \frac{1}{2}f^2\sin^2\theta\cos^2\theta \tag{3.9}$$

$$\frac{a}{r} \approx 1 + f\cos^2\theta + f^2\cos^2\theta + \frac{1}{2}f^2\cos^2\theta - \frac{1}{2}f^2\cos^4\theta$$

$$\approx 1 + f\left(1 + \frac{3}{2}f\right)\cos^2\theta - \frac{1}{2}f^2\cos^4\theta \tag{3.10}$$

For some purposes it suffices to know the equation of the ellipticity only to first order in f. This is derived in Box 3.1.

3.2 The geopotential

The main component of gravity is the gravitational acceleration $\mathbf{a}_G$ towards the center of the Earth. This component varies with latitude because of the varying radius of the spheroid. The deviation from a spherical shape results from the deforming effect of Earth's rotation, which produces a centrifugal acceleration $\mathbf{a}_c$ directed perpendicular to and away from the axis of rotation (Fig. 3.2). This component is proportional to the distance from the rotation axis, so it also varies with latitude.

Gravity is the vector combination of the centrifugal and gravitational components, each of which is conservative and is the gradient of a scalar potential. The potential of gravity U_g at a point on Earth's surface, the geopotential, is the sum of the gravitational potential U_G and the centrifugal potential U_c at that point,

$$U_g = U_G + U_c \tag{3.11}$$

Box 3.1.   First-order equation of a slightly flattened spheroid

The equation in polar coordinates of an ellipse with semi-major axis a and ellipticity f is, from (3.8),

$$\frac{r}{a} = \frac{1-f}{1-f\sin^2\theta} \tag{1}$$

This equation can be expanded using the binomial theorem:

$$\frac{r}{a} = (1-f)(1-f\sin^2\theta)^{-1} \approx (1-f)(1+f\sin^2\theta + \cdots) \tag{2}$$

Because f is equal to $1/298.252$ (Table 2.1), the quantity f^2 is of the order of 10^{-5} and is for many purposes negligibly small. The binomial expansion may be curtailed to first order in f, giving

$$\frac{r}{a} = 1 - f + f\sin^2\theta = 1 - f(1-\sin^2\theta) \tag{3}$$

$$\frac{r}{a} \approx 1 - f\cos^2\theta \tag{4}$$

It is often convenient to express the elliptical polar section in terms of the Legendre polynomial $P_2(\cos\theta)$. Rearranging the equation for $P_2(\cos\theta)$ from Table 1.1 gives

$$\cos^2\theta = \frac{1}{3}(1 + 2P_2(\cos\theta)) \tag{5}$$

By substituting into (4) above, we get

$$\frac{r}{a} \approx 1 - \frac{f}{3} - \frac{2}{3}fP_2(\cos\theta) \tag{6}$$

Upon invoking the binomial expansion and ignoring terms of second order and higher in f, this reduces to

$$\frac{r}{a} \approx \left(1 - \frac{f}{3}\right)\left(1 - \frac{2}{3}fP_2(\cos\theta)\right) \tag{7}$$

Let R be the radius of a sphere with the same volume as the spheroid (Fig. 3.1). Then, omitting the factor $4\pi/3$ common to each volume, we have

$$R^3 = a^2c = a^3(1-f) \tag{8}$$

Taking the cube root and using the binomial expansion to first order gives

$$R = a(1-f)^{1/3} \approx a\left(1 - \frac{f}{3}\right) \tag{9}$$

Thus the equation for the radius of an elliptical polar section of the Earth in terms of the Legendre polynomial $P_2(\cos\theta)$, the flattening f, and the mean radius R of an equivalent sphere is

$$r = R\left(1 - \frac{2}{3}fP_2(\cos\theta)\right) \tag{10}$$

This is a useful first-order approximation to the shape of the Earth.

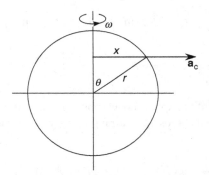

Fig. 3.2. Centrifugal acceleration $\mathbf{a}_c$ at co-latitude θ, directed perpendicular to and away from the axis of rotation.

3.2.1 Gravitational potential

To compute gravity on the reference spheroid it is necessary to determine the geopotential to second order in the small quantities that define it. Each of the quantities f, m, and J_2 is around 10^{-3} in size (Table 2.1), so their squares and products are around 10^{-6}. The gravitational potential (2.144) must be determined with the same definition, which means that it is inadequate to use only the terms up to J_2. If we assume that the mass distribution of the Earth is symmetric about the equator, the term J_3 can be omitted, but we need to include the term J_4 for an accurate description of the gravitational potential. Up to the term J_4 this becomes

$$U_G = -\frac{GE}{a}\left[\left(\frac{a}{r}\right) - J_2\left(\frac{a}{r}\right)^3 P_2(\cos\theta) - J_4\left(\frac{a}{r}\right)^5 P_4(\cos\theta)\right] \quad (3.12)$$

3.2.2 Centrifugal potential

The centrifugal acceleration is the gradient of the centrifugal potential U_c,

$$\mathbf{a}_c = -\nabla U_c \quad (3.13)$$

Let x be the perpendicular distance from the rotation axis to a point on the surface at latitude θ and let ω be the angular rate of rotation of the Earth (Fig. 3.2). The centrifugal acceleration is equal to $\omega^2 x$, so, for a constant rate of rotation, U_c varies only with x. Therefore

$$\omega^2 x = -\frac{\partial U_c}{\partial x} \quad (3.14)$$

Integrating both sides with respect to x gives

$$U_c = -\frac{1}{2}\omega^2 x^2 + U_0 \quad (3.15)$$

The potential is zero at the axis of rotation, where $x = 0$, and the constant of integration $U_c = 0$. The equation for the centrifugal potential in terms of the polar angle θ is

$$U_c = -\frac{1}{2}\omega^2 x^2 = -\frac{1}{2}\omega^2 r^2 \sin^2\theta \quad (3.16)$$

3.3 The equipotential surface of gravity

In order to compute gravity accurately on the reference ellipsoid it is necessary to develop the geopotential to second order in the small quantities f, m, and J_2, so we must use also the gravitational potential coefficient J_4 whose magnitude is around 10^{-6}. The geopotential consists of the sum of the gravitational and centrifugal potentials:

$$U_g = -\frac{GE}{a}\left[\left(\frac{a}{r}\right) - J_2\left(\frac{a}{r}\right)^3 P_2(\cos\theta) - J_4\left(\frac{a}{r}\right)^5 P_4(\cos\theta)\right]$$
$$- \frac{1}{2}\omega^2 a^2\left(\frac{r}{a}\right)^2 \sin^2\theta \quad (3.17)$$

Taking the centrifugal term inside the bracketed expression gives

$$U_g = -\frac{GE}{a}\left[\left(\frac{a}{r}\right) - J_2\left(\frac{a}{r}\right)^3 P_2(\cos\theta) - J_4\left(\frac{a}{r}\right)^5 P_4(\cos\theta) + \frac{1}{2}\left(\frac{\omega^2 a^3}{GE}\right)\left(\frac{r}{a}\right)^2 \sin^2\theta\right]$$

(3.18)

The geopotential involves the ratios a/r, $(a/r)^3$, and $(a/r)^5$, which we develop using (3.10). Note that the term in $(a/r)^3$ is multiplied by J_2 so it must be evaluated only to first order in f; the coefficient J_4 is itself of order 10^{-6}, so the ratio $(a/r)^5$ on the equipotential surface of gravity may be set equal to 1. Then

$$\left(\frac{a}{r}\right)^3 \approx 1 + 3\left(f\left(1 + \frac{3}{2}f\right)\cos^2\theta - \frac{1}{2}f^2\cos^4\theta\right) \approx 1 + 3f\cos^2\theta$$

(3.19)

For succinctness, let the last term inside the brackets in (3.18) be called Ψ. The ratio r/a is obtained from (3.8), thus

$$\Psi = \frac{1}{2}\left(\frac{\omega^2 a^3}{GE}\right)\left(\frac{r}{a}\right)^2 \sin^2\theta$$

$$= \frac{1}{2}\left(\frac{\omega^2 a^3(1-f)}{GE}\right)\frac{(1-f)\sin^2\theta}{(1-f\sin^2\theta)^2}$$

(3.20)

$$\Psi = \frac{1}{2}m\frac{(1-f)\sin^2\theta}{(1-f\sin^2\theta)^2}$$

(3.21)

Here m is the centrifugal acceleration ratio defined in Box 3.2, equation (3),

$$m = \frac{\omega^2 a^3(1-f)}{GE}$$

(3.22)

The denominator in (3.21) can be expanded using the binomial theorem; we need do so only to first order because of the factor m, which is similar in size to f. The centrifugal term Ψ becomes

$$\Psi \approx \frac{1}{2}m(1-f)(1 + 2f\sin^2\theta)\sin^2\theta$$

(3.23)

Multiplying, and retaining only the terms of first order in f, gives

$$\Psi = \frac{1}{2}m\sin^2\theta(1 - f + 2f\sin^2\theta)$$

(3.24)

In the equation for the geopotential, the centrifugal term must be combined with a term in $J_2 P_2(\cos\theta)$, which has the form $\cos^2\theta$, and a term in $J_4 P_4(\cos\theta)$, which

---

Box 3.2. **The acceleration ratio, *m***

The magnitudes of the gravitational and centrifugal components of gravity can be directly compared at the equator where the vectors are directly opposed to each other. The parameter *m* is defined as the ratio of the centrifugal acceleration at the equator to the gravitational acceleration at the equator:

$$m = \frac{\omega^2 a}{GE/a^2} = \frac{\omega^2 a^3}{GE} \tag{1}$$

The value of *m* defined in this way is $3.461\,391 \times 10^{-3} = 1/288.901$.

An alternative, commonly used definition of *m* is the ratio of the equatorial centrifugal acceleration to the gravitational acceleration on a sphere with the same volume as the spheroid. The volume of a spheroid with equatorial radius *a* and polar radius *c* is $(4\pi/3)a^2c$. The flattening *f* relates *a* and *c* so that $c = a(1 - f)$. Let the radius of a sphere with the same volume be *R*; its volume is $(4\pi/3)R^3$. On comparing the volumes and dropping the common numerical factor, we have

$$R^3 = a^2c = a^3(1 - f) \tag{2}$$

The alternative definition of the acceleration ratio *m* is then

$$m = \frac{\omega^2 R^3}{GE} = \frac{\omega^2 a^3 (1 - f)}{GE} \tag{3}$$

In this case the value of *m* is $3.449\,786 \times 10^{-3} = 1/289.873$.

---

contains terms in both $\cos^2\theta$ and $\cos^4\theta$ (see Table 1.2). It is advantageous to convert (3.24) to the same format:

$$\Psi = \frac{1}{2}m\left(1 - \cos^2\theta\right)\left(1 - f + 2f(1 - \cos^2\theta)\right)$$
$$= \frac{1}{2}m\left(1 + f - 2f\cos^2\theta - \cos^2\theta - f\cos^2\theta + 2f\cos^4\theta\right) \tag{3.25}$$

$$\Psi = \frac{1}{2}m\left(1 + f - (1 + 3f)\cos^2\theta + 2f\cos^4\theta\right) \tag{3.26}$$

Now we can return to (3.18). By writing the full expressions for $P_2(\cos\theta)$ and $P_4(\cos\theta)$ from Table 1.1, and the ratios a/r from (3.10) and $(a/r)^3$ from (3.19), and using (3.26) for the centrifugal term, we get the geopotential as a function of $\cos^2\theta$ and $\cos^4\theta$:

$$U_g = -\frac{GE}{a}\begin{bmatrix} 1 + f(1 + \frac{3}{2}f)\cos^2\theta - \frac{1}{2}f^2\cos^4\theta \\ - J_2(-1 + 3(1-f)\cos^2\theta + 9f\cos^4\theta)/2 \\ - J_4(3 - 30\cos^2\theta + 35\cos^4\theta)/8 \\ + \frac{1}{2}m(1 + f - (1+3f)\cos^2\theta + 2f\cos^4\theta) \end{bmatrix} \tag{3.27}$$

After gathering terms to get the coefficients that multiply $\cos^2\theta$ and $\cos^4\theta$, we get the final expression for the geopotential:

$$U_g = -\frac{GE}{a}\begin{bmatrix} 1 + f + \frac{1}{2}m + \frac{1}{2}J_2 - \frac{3}{8}J_4 \\ + \left(f + \frac{3}{2}f^2 - \frac{1}{2}m - \frac{3}{2}fm - \frac{3}{2}(1-f)J_2 + \frac{15}{4}J_4\right)\cos^2\theta \\ - \left(\frac{1}{2}f^2 - mf + \frac{9}{2}fJ_2 + \frac{35}{8}J_4\right)\cos^4\theta \end{bmatrix} \tag{3.28}$$

3.3.1 Relationship of J_2, J_4, f, and m

By definition, the geopotential must be constant on the equipotential surface. However, the potential in (3.28) can vary with polar angle through the terms in $\cos^2\theta$ and $\cos^4\theta$. This apparent contradiction implies that the coefficients of these terms must be zero, i.e.,

$$f - \frac{1}{2}m + \frac{3}{2}f^2 - \frac{3}{2}fm - \frac{3}{2}(1-f)J_2 + \frac{15}{4}J_4 = 0 \tag{3.29}$$

$$\frac{1}{2}f^2 - mf + \frac{9}{2}fJ_2 + \frac{35}{8}J_4 = 0 \tag{3.30}$$

Since J_4 is much smaller than J_2, we can neglect it initially and write (3.29) to first order:

$$f - \frac{1}{2}m - \frac{3}{2}J_2 = 0 \tag{3.31}$$

$$J_2 = \frac{1}{3}(2f - m) \tag{3.32}$$

This value for J_2 is now inserted into (3.30) to obtain a second-order equation for J_4:

$$J_4 = \frac{8}{35}\left(-\frac{1}{2}f^2 + mf - \frac{9}{2}f\left(\frac{2}{3}f - \frac{1}{3}m\right)\right)$$
$$= \frac{4}{7}fm - \frac{4}{5}f^2 \tag{3.33}$$

By inserting this expression back into (3.29) we eliminate J_4 and get an equation for J_2:

$$(1-f)J_2 = \frac{2}{3}f - \frac{1}{3}m + f^2 - fm + \frac{5}{2}\left(\frac{4}{7}fm - \frac{4}{5}f^2\right) \tag{3.34}$$

$$(1-f)J_2 = \frac{2}{3}f - \frac{1}{3}m - f^2 + \frac{3}{7}fm \tag{3.35}$$

Applying the binomial theorem to first order in f gives

$$J_2 = \left(\frac{2}{3}f - \frac{1}{3}m - f^2 + \frac{3}{7}fm\right)(1 + f + \cdots) \tag{3.36}$$

After multiplying and tidying up the terms, we get a second-order equation for J_2:

$$J_2 = \frac{1}{3}\left(2f - m - f^2 + \frac{2}{7}fm\right) \tag{3.37}$$

3.3.2 Inferred increase of density with depth in the Earth

In Section 2.5.2 the dynamic form factor J_2 is expressed in terms of the principal moments of inertia. We can replace the equatorial radius a by the mean radius R, so that to first order

$$J_2 = \frac{C-A}{Ea^2} \approx \frac{C-A}{ER^2} \tag{3.38}$$

By combining this result with (3.32) we obtain a relationship among the difference in the principal moments of inertia, the flattening responsible for the difference, and the centrifugal acceleration that causes the deformation:

$$\frac{C-A}{ER^2} = \frac{1}{3}(2f - m) \tag{3.39}$$

Equation (3.39) allows us to make an inference about the distribution of mass inside the Earth. The Sun and Moon exert torques on the spheroidal shape of the Earth that cause the rotation axis to precess about the pole to the ecliptic plane,

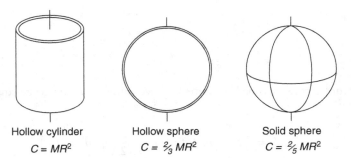

Fig. 3.3. Moments of inertia of a hollow cylinder, hollow sphere, and uniform solid sphere about an axis of symmetry.

which is manifest in the precession of the equinoxes (see Section 5.3). The rate of precession is determined by the *dynamic ellipticity H*, defined as

$$H = \frac{C - (A + B)/2}{C} \approx \frac{C - A}{C} \tag{3.40}$$

The value of H is known quite accurately from astronomic observations. H is a very small quantity of the same order as f and m (Table 2.1). Rewriting (3.39) gives

$$\left(\frac{C - A}{C}\right)\frac{C}{ER^2} = \frac{1}{3}(2f - m) \tag{3.41}$$

$$\frac{C}{ER^2} = \frac{1}{3}\left(\frac{2f - m}{H}\right) \approx \frac{1}{3} \tag{3.42}$$

$$C \approx \frac{1}{3}ER^2 \tag{3.43}$$

Figure 3.3 shows the moments of inertia of some standard objects about an axis of symmetry. With increasing distribution of the mass of the object closer to its center, the factor preceding the product MR^2 decreases from 1 for an open-ended hollow cylinder to 0.67 for a hollow spherical shell and 0.4 for a homogeneous solid sphere. The numerical factor is 0.33 for the Earth, indicating that the density of the Earth is not uniform but increases towards its center, i.e., the density of the Earth increases with depth.

3.4 Gravity on the reference spheroid

The reference figure for standard calculations of gravity at a particular latitude is the spheroid, or ellipsoid, of revolution. The acceleration due to gravity on the reference spheroid has both a radial component g_r and a polar component g_θ,

$$\mathbf{g} = g_r\mathbf{e}_r + g_\theta\mathbf{e}_\theta \tag{3.44}$$

The polar component g_θ is much smaller than the radial component g_r, but it has important effects. It deflects the vertical from the radial direction at every point on the Earth, except at the poles and on the equator. This deflection results in a difference between geocentric and geographic latitude; the maximum difference is less than $0.2°$, but this has a large effect on measurements of gravity. The polar component cannot be neglected, since this would be akin to assuming that gravity acts in a radial direction at all points. To determine the theoretical gravity on the reference spheroid we must combine expressions for the radial and polar components:

$$g = \left((g_r)^2 + (g_\theta)^2\right)^{1/2} \approx g_r\left(1 + \frac{1}{2}\left(\frac{g_\theta}{g_r}\right)^2\right) \tag{3.45}$$

As we will see, the polar component g_θ is of order f, so its effect on gravity is proportional to f^2. To determine the variation of gravity on the reference ellipsoid we will have to evaluate the radial component to second order as well. This makes it necessary to express the shape of the spheroid and the geopotential to second order in the small quantities f, m, and J_2. We must also use an expression for the gravitational potential up to the coefficient J_4, which is about the same size as the squares and products of these parameters.

3.4.1 Polar component of gravity

The polar component of gravity on the reference ellipsoid is the gradient of the geopotential in the direction of increasing polar angle θ,

$$g_\theta = -\frac{1}{r}\frac{\partial}{\partial\theta}\left\{-\frac{GE}{a}\left[\left(\frac{a}{r}\right) - J_2\left(\frac{a}{r}\right)^3 P_2(\cos\theta) - J_4\left(\frac{a}{r}\right)^5 P_4(\cos\theta)\right] - \frac{1}{2}\omega^2 r^2 \sin^2\theta\right\} \tag{3.46}$$

The first term is independent of θ and drops out of the differentiation. We can take the centrifugal term inside the square brackets and use the definition of the centrifugal ratio m as in (3.22):

$$g_\theta = -\frac{GE}{a^2}\left[J_2\left(\frac{a}{r}\right)^4\frac{\partial}{\partial\theta}P_2(\cos\theta) + J_4\left(\frac{a}{r}\right)^6\frac{\partial}{\partial\theta}P_4(\cos\theta)\right.$$
$$\left. - \frac{1}{2}\left(\frac{m}{1-f}\right)\left(\frac{r}{a}\right)\frac{\partial}{\partial\theta}\sin^2\theta\right] \tag{3.47}$$

The Legendre polynomials $P_2(\cos\theta)$ and $P_4(\cos\theta)$ are listed in Table 1.1. Differentiating them with respect to θ gives

$$\frac{\partial}{\partial\theta}P_2(\cos\theta) = \frac{\partial}{\partial\theta}\left(\frac{3\cos^2\theta - 1}{2}\right) = -3\cos\theta\sin\theta \qquad (3.48)$$

$$\frac{\partial}{\partial\theta}P_4(\cos\theta) = \frac{\partial}{\partial\theta}\left(\frac{35\cos^4\theta - 30\cos^2\theta + 3}{8}\right)$$
$$= -\frac{5}{2}\cos\theta\sin\theta(7\cos^2\theta - 3) \qquad (3.49)$$

On substituting these into (3.47) and simplifying, we obtain

$$g_\theta = \frac{GE}{a^2}\sin\theta\cos\theta\left[3J_2\left(\frac{a}{r}\right)^4 + \frac{5}{2}J_4\left(\frac{a}{r}\right)^6(7\cos^2\theta - 3) + \frac{m}{1-f}\left(\frac{r}{a}\right)\right]$$
$$(3.50)$$

As explained above, we need to evaluate g_θ only to first order in f, so terms with J_4 and the products fJ_2 and fm may be neglected. The ratios $(a/r)^4$, $(a/r)^6$, and r/a may be set effectively equal to 1. We define

$$g_0 = \frac{GE}{a^2} \qquad (3.51)$$

The polar component of gravity on the reference ellipsoid is therefore given to first order by

$$g_\theta \approx g_0(3J_2 + m)\sin\theta\cos\theta \qquad (3.52)$$

Now we recall the relationship among J_2, f, and m established in (3.32) and substitute for J_2, which gives the first-order expression

$$g_\theta \approx g_0 f \sin(2\theta) \qquad (3.53)$$

Note that g_θ is positive for $\theta \leq 90°$ and negative for $90° \leq \theta \leq 180°$, i.e., in each hemisphere g_θ acts in the direction from the pole to the equator.

3.4.2 Radial component of gravity

The radial component of gravity on the reference ellipsoid is obtained from the gradient of the geopotential with respect to the radius r:

$$g_r = -\frac{\partial}{\partial r}\left\{-\frac{GE}{a}\left[\left(\frac{a}{r}\right) - J_2\left(\frac{a}{r}\right)^3 P_2(\cos\theta) - J_4\left(\frac{a}{r}\right)^5 P_4(\cos\theta)\right.\right.$$
$$\left.\left. + \frac{1}{2}\left(\frac{\omega^2 a^3}{GE}\right)\left(\frac{r}{a}\right)^2\sin^2\theta\right]\right\} \qquad (3.54)$$

$$g_r = -\frac{GE}{a^2}\left[\left(\frac{a}{r}\right)^2 - 3J_2\left(\frac{a}{r}\right)^4 P_2(\cos\theta) - 5J_4\left(\frac{a}{r}\right)^6 P_4(\cos\theta)\right.$$
$$\left. - \left(\frac{m}{1-f}\right)\left(\frac{r}{a}\right)\sin^2\theta\right] \tag{3.55}$$

To simplify this cumbersome evaluation somewhat, we examine the four terms inside the square brackets individually. We write g_0 as in (3.51):

$$g_r = -g_0[T_1 + T_2 + T_3 + T_4] \tag{3.56}$$

For term T_1, using the ratio a/r defined in (3.10), and neglecting terms of higher order than f^2, the first term in square brackets is

$$\left(\frac{a}{r}\right)^2 = 1 + 2f\left(\left(1+\frac{3}{2}f\right)\cos^2\theta - \frac{1}{2}f\cos^4\theta\right)$$
$$+ f^2\cos^4\theta\left(\left(1+\frac{3}{2}f\right) - \frac{1}{2}f\cos^2\theta\right)^2 \tag{3.57}$$

Thus

$$T_1 \approx 1 + (2f + 3f^2)\cos^2\theta \tag{3.58}$$

For term T_2, the term in $(a/r)^4$ is multiplied by J_2, so we need only expand it to order f:

$$\left(\frac{a}{r}\right)^4 \approx 1 + 4f\left(\left(1+\frac{3}{2}f\right)\cos^2\theta - \frac{1}{2}f\cos^4\theta\right) \tag{3.59}$$

$$\left(\frac{a}{r}\right)^4 \approx 1 + 4f\cos^2\theta \tag{3.60}$$

Using the expansion of the Legendre polynomial $P_2(\cos\theta)$ given in Table 1.1,

$$T_2 = -3J_2(1 + 4f\cos^2\theta)P_2(\cos\theta) = -\frac{3}{2}J_2(1 + 4f\cos^2\theta)(3\cos^2\theta - 1) \tag{3.61}$$

$$T_2 \approx \frac{3}{2}J_2 - 3\left(\frac{3}{2} - 2f\right)J_2\cos^2\theta - 18fJ_2\cos^4\theta \tag{3.62}$$

For term T_3, the term in $(a/r)^6$ is multiplied by J_4, which is of order 10^{-6}, so we can neglect products of J_4 with f. Effectively we can set $(a/r)^6$ equal to 1. Using the expansion of $P_4(\cos\theta)$,

$$T_3 \approx -5J_4P_4(\cos\theta) \approx -\frac{5}{8}J_4(3 - 30\cos^2\theta + 35\cos^4\theta) \tag{3.63}$$

For term T_4, the ratio r/a is given by (3.8), and to second order this term is

$$T_4 \approx -\left(\frac{m}{1-f}\right)\sin^2\theta\,\frac{1-f}{1-f\sin^2\theta} \approx -m\sin^2\theta(1+f\sin^2\theta) \qquad (3.64)$$

On converting the sines to cosines for compatibility with the other terms we obtain

$$T_4 \approx -m(1+f) + m(1+2f)\cos^2\theta - mf\cos^4\theta \qquad (3.65)$$

Now we can insert these four terms into (3.56):

$$g_r = -g_0 \begin{bmatrix} 1 + f(2+3f)\cos^2\theta \\[4pt] + \tfrac{3}{2}J_2 - 3\left(\tfrac{3}{2} - 2f\right)J_2\cos^2\theta - 18fJ_2\cos^4\theta \\[4pt] - \tfrac{5}{8}J_4\left(3 - 30\cos^2\theta + 35\cos^4\theta\right) \\[4pt] - m(1+f) + m(1+2f)\cos^2\theta - mf\cos^4\theta \end{bmatrix} \qquad (3.66)$$

After gathering terms to form coefficients of $\cos^2\theta$ and $\cos^4\theta$, we have

$$g_r = -g_0 \begin{bmatrix} 1 + \tfrac{3}{2}J_2 - \tfrac{15}{8}J_4 - m(1+f) \\[4pt] + \left(f(2+3f) - 3\left(\tfrac{3}{2} - 2f\right)J_2 + \tfrac{75}{4}J_4 + m(1+2f)\right)\cos^2\theta \\[4pt] - \left(mf + 18fJ_2 + \tfrac{175}{8}J_4\right)\cos^4\theta \end{bmatrix}$$

$$ (3.67)$$

J_2 and J_4 can be replaced by expressions in f and m, as in (3.37) and (3.33), respectively. After expanding and grouping the terms, the radial gravity component becomes

$$g_r = -g_0 \begin{bmatrix} 1 + f - \tfrac{3}{2}m + f^2 - \tfrac{27}{14}fm \\[4pt] + \left(\tfrac{5}{2}m - f - \tfrac{13}{2}f^2 + \tfrac{72}{7}fm\right)\cos^2\theta \\[4pt] - \left(\tfrac{15}{2}fm - \tfrac{11}{2}f^2\right)\cos^4\theta \end{bmatrix} \qquad (3.68)$$

3.4.3 Variation of reference gravity with geocentric latitude

Instead of using the polar angle θ to describe position on the reference ellipsoid, it is customary to use the latitude. The geocentric latitude λ_c is the complement of θ, so $\cos\theta = \sin\lambda_c$, $\cos^2\theta = \sin^2\lambda_c$, and

$$\cos^4\theta = \sin^4\lambda_c = \sin^2\lambda_c\left(1 - \cos^2\lambda_c\right) = \sin^2\lambda_c - \frac{1}{4}\sin^2(2\lambda_c) \qquad (3.69)$$

On substituting this change, the radial component of gravity on the spheroid as a function of geocentric latitude is

$$g_r = -g_0 \begin{bmatrix} 1 + f - \frac{3}{2}m + f^2 - \frac{27}{14}fm \\ + \left(\frac{5}{2}m - f - f^2 + \frac{39}{14}fm\right)\sin^2\lambda_c \\ + \frac{1}{8}f(15m - 11f)\sin^2(2\lambda_c) \end{bmatrix} \tag{3.70}$$

Note that the polar component g_θ (see (3.53)) referred to geocentric latitude is unaltered:

$$g_\theta \approx g_0 f\sin(2\theta) = g_0 f\sin(2\lambda_c) \tag{3.71}$$

Gravity on the reference figure of the Earth acts normal to the ellipsoidal equipotential surface. It is computed by combining the radial and polar components as in (3.45):

$$g = g_r\left(1 + \frac{1}{2}\left(\frac{g_\theta}{g_r}\right)^2\right) = g_r\left(1 + \frac{1}{2}f^2\sin^2(2\lambda_c)\left(1 + f - \frac{3}{2}m + \cdots\right)^{-2}\right) \tag{3.72}$$

$$g \approx g_r\left(1 + \frac{1}{2}f^2\sin^2(2\lambda_c)\right) \tag{3.73}$$

Thus the polar component affects only the $\sin^2(2\lambda_c)$ term in (3.70), and gravity on the reference ellipsoid is given by

$$g = -g_0 \begin{bmatrix} 1 + f - \frac{3}{2}m + f^2 - \frac{27}{14}fm \\ + \left(\frac{5}{2}m - f - f^2 + \frac{39}{14}fm\right)\sin^2\lambda_c \\ + \frac{1}{8}f(15m - 7f)\sin^2(2\lambda_c) \end{bmatrix} \tag{3.74}$$

Let the value of gravity at the equator, where $\sin\lambda_c = \sin(2\lambda_c) = 0$, be

$$g_e = -g_0\left(1 + f - \frac{3}{2}m + f^2 - \frac{27}{14}fm\right) \tag{3.75}$$

Taking this out of the bracketed expression and using the binomial expansion to first order in f gives

$$g \approx g_e\left\{1 + A\sin^2\lambda_c + \frac{1}{8}f(15m - 7f)\sin^2(2\lambda_c)\right\}\left(1 - \left(f - \frac{3}{2}m + f^2 - \frac{27}{14}fm\right)\right) \tag{3.76}$$

where, for succinctness, $A = \frac{5}{2}m - f - f^2 + \frac{39}{14}fm$. The coefficient of $\sin^2(2\lambda_c)$ is already of second order, so, when we multiply the terms, only the coefficient A of $\sin^2\lambda_c$ is affected. It expands to

$$\left(\frac{5}{2}m - f - f^2 + \frac{39}{14}fm\right)\left(1 - f + \frac{3}{2}m - f^2 + \frac{27}{14}fm\right)$$

$$= \frac{5}{2}m - f - f^2 + \frac{39}{14}fm - \frac{5}{2}fm + f^2 + \frac{15}{4}m^2 - \frac{3}{2}fm$$

$$= \frac{5}{2}m - f + \frac{15}{4}m^2 - \frac{17}{14}fm \tag{3.77}$$

The final expression for the variation of gravity with geocentric latitude is

$$g = g_e\left\{1 + \left(\frac{5}{2}m - f + \frac{15}{4}m^2 - \frac{17}{14}fm\right)\sin^2\lambda_c + \frac{1}{8}f(15m - 7f)\sin^2(2\lambda_c)\right\} \tag{3.78}$$

3.4.4 Clairaut's formula

The value of gravity at the poles, g_p, is found by setting $\lambda_c = \pi/2 = 90°$. To first order

$$g_p = g_e\left\{1 + \left(\frac{5}{2}m - f\right)\right\} \tag{3.79}$$

Rearranging this equation gives

$$\frac{g_p - g_e}{g_e} = \frac{5}{2}m - f \tag{3.80}$$

This is the Clairaut formula for the difference between the gravity at the pole and that at the equator, attributed to a French mathematician and astronomer, Alexis Claude de Clairaut (1713–1765).

3.5 Geocentric and geographic latitude

The latitude in the above formulae is the geocentric latitude λ_c defined by the radius from the Earth's center to the point on the ellipsoid. However, the latitude in common use is the geographic (or geodetic) latitude λ defined by the vertical direction, which is normal to the surface of the reference ellipsoid and does not pass through the Earth's center (Fig. 3.4). There is a simple relationship between the geocentric and geographic latitudes.

Let P be a point on the ellipsoid with geocentric latitude λ_c and geographic latitude λ (Fig. 3.5(a)). The angle between the radial and vertical directions at P is $\lambda - \lambda_c$. The horizontal direction PH and the direction PN normal to the radius

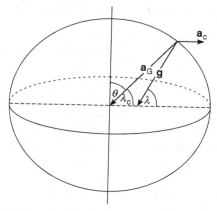

Fig. 3.4. Comparison of geocentric latitude λ_c, defined by the radius of the ellipsoidal Earth, and geographic (or geodetic) latitude λ, defined by the normal direction to the surface of the ellipsoid. After Lowrie (2007).

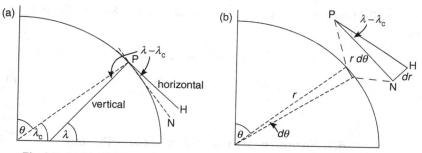

Fig. 3.5. (a) The difference $(\lambda - \lambda_c)$ between geographic latitude λ and geocentric latitude λ_c is the same as the angle between the horizontal and a plane perpendicular to the radius. (b) Details of the construction of a small triangle whose sides PN and PH contain the angle $(\lambda - \lambda_c)$.

at P form the same angle. Consider a small increase $d\theta$ in the polar angle for the point P. The radius to the surface increases by a small amount dr, and there is an angular displacement $r\,d\theta$ perpendicular to the radius. These increments displace the intersection of the radius with the surface along the ellipsoid. The three displacements form a small triangle PNH (Fig. 3.5(b)), whose sides PN and PH contain the angle $(\lambda - \lambda_c)$. In the triangle PNH

$$\tan(\lambda - \lambda_c) = \frac{dr}{r\,d\theta} \tag{3.81}$$

On differentiating the equation of the ellipsoid (3.8) we have to first order in f

$$\begin{aligned}
\frac{1}{r}\frac{dr}{d\theta} &= \frac{a}{r}\frac{d}{d\theta}\frac{1-f}{1-f\sin^2\theta} \\
&= \frac{af(2\sin\theta\cos\theta)}{r\left(1-f\sin^2\theta\right)^2} \\
&\approx f\sin(2\theta)
\end{aligned} \tag{3.82}$$

Because θ is the complement of λ_c, we can replace $\sin(2\theta)$ by $\sin(2\lambda_c)$ and obtain the result

$$\tan(\lambda - \lambda_c) = f\sin(2\lambda_c) \tag{3.83}$$

The difference $\delta\lambda = \lambda - \lambda_c$ is very small, because the tangent of the angle is less than f,

$$\delta\lambda = \lambda - \lambda_c \leq \tan^{-1}(f) \leq 0.19° \tag{3.84}$$

The small difference allows us to replace the tangent in (3.83) with the angle (in radians), so that

$$\lambda = \lambda_c + f\sin(2\lambda_c) \tag{3.85}$$

$$\delta\lambda = \lambda - \lambda_c = f\sin(2\lambda_c) \tag{3.86}$$

3.5.1 Normal gravity on the reference ellipsoid

Measurements of gravity must be corrected for various factors, such as the latitude of the measurement site, its altitude with respect to the reference ellipsoid, and the surrounding topography. The corrected value must then be compared with the theoretical value for the geographic latitude of the observation. The gravity formula in (3.78) gives the variation of gravity with *geocentric* latitude. This must now be converted to a form that depends on geographic latitude, which requires finding expressions for $\sin^2\lambda_c$ and $\sin^2(2\lambda_c)$ in terms of λ.

The gravity formula in (3.78) can be written

$$g_n = g_e(1 + b_1\sin^2\lambda_c + b_2\sin^2(2\lambda_c)) \tag{3.87}$$

On comparing (3.87) with (3.78), we note that the constant b_1 contains terms of both first and second order in f and m, whereas b_2 is entirely of second order. This allows us to simplify the conversions.

From (3.86) we have $\lambda_c = \lambda - \delta\lambda$ and, because $\delta\lambda$ is a very small angle, we can make the approximations $\sin(\delta\lambda) \approx \delta\lambda$ and $\cos(\delta\lambda) \approx 1$. The expressions for $\sin\lambda_c$ and $\cos\lambda_c$ reduce to

$$\sin\lambda_c = \sin(\lambda - \delta\lambda) = \sin\lambda\cos(\delta\lambda) - \cos\lambda\sin(\delta\lambda) \approx \sin\lambda - \delta\lambda\cos\lambda$$
(3.88)

$$\cos\lambda_c = \cos(\lambda - \delta\lambda) = \cos\lambda\cos(\delta\lambda) + \sin\lambda\sin(\delta\lambda) \approx \cos\lambda + \delta\lambda\sin\lambda$$
(3.89)

The gravity formula contains the term $\sin^2\lambda_c$, which we can now write as

$$\sin^2\lambda_c \approx \sin^2\lambda - 2\delta\lambda\sin\lambda\cos\lambda \approx \sin^2\lambda - \delta\lambda\sin(2\lambda) \qquad (3.90)$$

Next, we combine (3.88) and (3.89) to get an expression for $\sin^2(2\lambda_c)$, which is, to first order in $\delta\lambda$,

$$\begin{aligned}
\sin(2\lambda_c) &= 2(\sin\lambda - \delta\lambda\cos\lambda)(\cos\lambda + \delta\lambda\sin\lambda) \\
&= 2\sin\lambda\cos\lambda - 2\delta\lambda(\cos^2\lambda - \sin^2\lambda) - 2(\delta\lambda)^2\sin\lambda\cos\lambda \quad (3.91) \\
&\approx \sin(2\lambda) - 2\delta\lambda\cos(2\lambda)
\end{aligned}$$

Squaring, and again neglecting the term in $(\delta\lambda)^2$, gives

$$\sin^2(2\lambda_c) \approx \sin^2(2\lambda) - 4\delta\lambda\sin(2\lambda)\cos(2\lambda) \approx \sin^2(2\lambda) - 2\delta\lambda\sin(4\lambda)$$
(3.92)

In the gravity formula (3.87) this term is multiplied by the constant b_2, which is of second order in f and m. Thus, neglecting the small product $b_2\,\delta\lambda$,

$$b_2\sin^2(2\lambda_c) \approx b_2\sin^2(2\lambda) - 2b_2\,\delta\lambda\sin(4\lambda) \approx b_2\sin^2(2\lambda) \qquad (3.93)$$

Equation (3.91) allows us to rewrite $\delta\lambda$ in (3.86),

$$\delta\lambda = f\sin(2\lambda_c) = f\sin(2\lambda) - (f\delta\lambda)\cos(2\lambda) \approx f\sin(2\lambda) \qquad (3.94)$$

Upon inserting this into (3.90) we get

$$\sin^2\lambda_c \approx \sin^2\lambda - f\sin^2(2\lambda) \qquad (3.95)$$

Substituting (3.93) and (3.95) into (3.87) gives the gravity formula for geographic latitude λ:

$$g_n = g_e\big(1 + b_1\big(\sin^2\lambda - f\sin^2(2\lambda)\big) + b_2\sin^2(2\lambda)\big) \qquad (3.96)$$

$$g_n = g_e\big(1 + b_1\sin^2\lambda + (b_2 - fb_1)\sin^2(2\lambda)\big) \qquad (3.97)$$

The coefficient of $\sin^2\lambda$ is the same as that of $\sin^2\lambda_c$ in the gravity formula (3.87) for geocentric latitude, but the coefficient of $\sin^2(2\lambda)$ is modified to

$$
\begin{aligned}
b_2 - fb_1 &= \frac{1}{8}f(15m - 7f) - f\left(\frac{5}{2}m - f + \frac{15}{4}m^2 - \frac{17}{14}fm\right) \\
&= \frac{1}{8}f(f - 5m)
\end{aligned}
\tag{3.98}
$$

On replacing b_1 and b_2 by the corresponding expressions in (3.78), we get the *normal gravity formula*

$$
g_n = g_e\left(1 + \beta_1 \sin^2\lambda + \beta_2 \sin^2(2\lambda)\right)
\tag{3.99}
$$

in which g_n is the normal gravity at geographic latitude λ on the International Reference Ellipsoid, g_e is its value at the equator, and β_1 and β_2 are small constants, given by

$$
\begin{aligned}
\beta_1 &= \frac{5}{2}m - f + \frac{15}{4}m^2 - \frac{17}{14}fm \\
\beta_2 &= \frac{1}{8}\left(f^2 - 5fm\right)
\end{aligned}
\tag{3.100}
$$

From (3.51) and (3.75) the value of gravity on the equator is given by

$$
g_e = -\frac{GE}{a^2}\left(1 + f - \frac{3}{2}m + f^2 - \frac{27}{14}fm\right)
$$

3.6 The geoid

The real surface of the Earth is irregular and cannot be described by a simple geometric form. It is replaced by a smooth equipotential surface of gravity, chosen so that it agrees with mean sea-level far from land. This surface is called the geoid. The distribution of density in the Earth's crust is complex, with local mass anomalies that influence the geoid and cause it to undulate about a mean shape. The mathematical reference figure for the Earth is a spheroid that has the same volume and the same potential as the geoid.

A local excess of mass deflects the direction of a plumb-line towards it and at the same time increases the local value of gravity. In order to maintain a constant potential, the equipotential surface must bulge upwards over the excess mass. The shape of the bulge is determined by the condition that the equipotential must lie normal to the direction of gravity and hence to the plumb-line. The mass excess elevates the geoid above the spheroid (Fig. 3.6); conversely, a mass

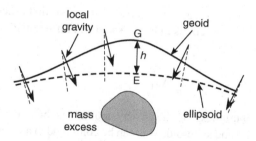

Fig. 3.6. Elevation of the geoid above the reference ellipsoid due to an excess of mass below the ellipsoid, and related local deflections of the direction of gravity. After Lowrie (2007).

deficit depresses the geoid below the spheroid. The undulations of the geoid with respect to the spheroid correlate with the gravity anomalies caused by the inhomogeneity of density. The height of the geoid relative to the spheroid may be calculated from an analysis of these gravity anomalies.

3.6.1 The potential of a geoid undulation

Let E be a point on the reference ellipsoid (idealized gravity equipotential) over an anomalous mass. The effect of this mass is to raise the geoid (true gravity equipotential) so that the point G corresponding to E is at a height h above the ellipsoid (Fig. 3.6). The work done against gravity g changes the potential. If the displacement h is small, the additional potential W due to the excess mass is simply $W = gh$. Thus the height of the geoid above the spheroid is

$$h = \frac{W}{g} \qquad (3.101)$$

Gravity observations are first corrected for local topography and transient tidal effects. The corrected value is then reduced to the reference surface by compensating for the altitude of the measurement station. A gravity anomaly is computed by subtracting the theoretical gravity for the latitude of the measurement station. However, altitudes are specified relative to mean sea-level, so the altitude adjustment reduces the gravity value to the geoid rather than the ellipsoid. The gravity anomaly after corrections and reduction is specified at the point G on the geoid, but the reference value is computed for the point E on the ellipsoid (Fig. 3.6). The height difference corresponds to the geoid undulation, which must be taken into account in an accurate gravity survey.

The gravity anomaly Δg at the point G arises from two superposed effects. The main effect is the gravitational attraction of the additional mass. This causes a vertical gravity anomaly Δg_1 that can be calculated to first order by assuming the vertical and radial directions to be the same and differentiating the potential W with respect to r,

$$\Delta g_1 = -\frac{\partial W}{\partial r} \tag{3.102}$$

The second contribution Δg_2 to the gravity anomaly is the effect of the distance h between the geoid and spheroid. This can be computed in an analogous way to the gravity free-air correction:

$$\Delta g_2 = h\frac{\partial g}{\partial r} \tag{3.103}$$

$$\frac{\partial g}{\partial r} = \frac{\partial}{\partial r}\left(-\frac{GE}{r^2}\right) = -2\frac{g}{r} \tag{3.104}$$

On combining the two contributions, we get for the gravity anomaly of the anomalous mass

$$\Delta g = \Delta g_1 + \Delta g_2 = -\left(\frac{\partial W}{\partial r} + 2\frac{W}{r}\right) \tag{3.105}$$

The geoid undulations h are much smaller than the Earth's radius R, so it is unimportant if this expression is evaluated on the spherical Earth rather than on the actual spheroid. We can conveniently use the surface of the sphere $r = R$, in which case

$$\Delta g = -\left(\frac{1}{r^2}\frac{\partial}{\partial r}\left(r^2 W\right)\right)_{r=R} \tag{3.106}$$

3.6.2 Stokes' formula for the height of the geoid

Suppose the height of the geoid is to be determined at a point P from gravity anomalies on the Earth's surface $r = R$. Let the spherical coordinates be defined relative to a radial axis through the point P. For a point Q where gravity was measured, θ is the polar angle relative to P and ϕ is the azimuth of Q on a circle around P. The gravity anomalies on the spherical surface can then be expressed as a sum of spherical harmonic functions, $Y_n^m(\theta, \phi)$ (see Section 1.16):

$$\Delta g(\theta, \phi) = \sum_{n=0}^{\infty} \sum_{m=0}^{n} g_n^m Y_n^m(\theta, \phi) \tag{3.107}$$

Also, the potential W of the excess mass must be a solution of Laplace's equation, so we can write

$$W = \sum_{n=0}^{\infty} \sum_{m=0}^{n} \frac{B_n^m Y_n^m(\theta, \phi)}{r^{n+1}} \tag{3.108}$$

Multiplying by r^2 gives

$$r^2 W = \sum_{n=0}^{\infty} \sum_{m=0}^{n} \frac{B_n^m Y_n^m(\theta, \phi)}{r^{n-1}} \tag{3.109}$$

Differentiating with respect to r gives

$$-\frac{\partial}{\partial r}(r^2 W) = \sum_{n=0}^{\infty} \sum_{m=0}^{n} (n-1) \frac{B_n^m Y_n^m(\theta, \phi)}{r^n} \tag{3.110}$$

Upon inserting this expression into (3.106) and evaluating on the surface $r = R$, we have

$$\Delta g(\theta, \phi) = \sum_{n=0}^{\infty} \sum_{m=0}^{n} (n-1) \frac{B_n^m Y_n^m(\theta, \phi)}{R^{n+2}} \tag{3.111}$$

Note that there is no term for $n = 1$ in this sum; also, the term for $n = 0$ is a constant, which may be considered part of the overall potential, but is not of interest for the anomalies. Thus the summation begins at $n = 2$. On comparing the coefficients of $Y_n^m(\theta, \phi)$ in (3.107) and (3.111), we have

$$\Delta g_n^m = (n-1) \frac{B_n^m}{R^{n+2}} \tag{3.112}$$

$$B_n^m = \frac{R^{n+2}}{n-1} g_n^m \tag{3.113}$$

This expression can now be substituted into (3.108) for the potential,

$$W = R \sum_{n=2}^{\infty} \sum_{m=0}^{n} \frac{1}{n-1} \left(\frac{R}{r}\right)^{n+1} \Delta g_n^m Y_n^m(\theta, \phi) \tag{3.114}$$

Computation of the height of the geoid is simplified by introducing a *zonal approximation*. The distribution of gravity anomalies $Y_n^m(\theta, \phi)$ is replaced by zonal harmonics, which are essentially the zeroth-order Legendre polynomials $P_n(\cos \theta)$. Effectively, the gravity anomalies at co-latitude θ are summed over longitude ϕ. Compared with (3.107), we make the replacement

$$\Delta\bar{g}_n P_n(\cos\theta) = \sum_{m=0}^{n} g_n^m Y_n^m(\theta,\phi) \qquad (3.115)$$

As a result the gravity anomalies on the surface of the sphere are now represented by

$$\Delta g(\theta,\phi) = \sum_{n=2}^{\infty} \Delta\bar{g}_n P_n(\cos\theta) \qquad (3.116)$$

In order to make use of the orthogonal properties of the Legendre polynomials (see Section 1.15), we multiply both sides by $P_n(\cos\theta)$ and integrate over the surface of the unit sphere. The element of surface area on the unit sphere (radius $r = 1$) is $d\Omega = \sin\theta\, d\theta\, d\phi$ (Box 1.3) and the limits of integration are $0 \le \theta \le \pi$ and $0 \le \phi \le 2\pi$. The integral is

$$\iint_S \Delta g(\theta,\phi) P_n(\cos\theta) d\Omega = \sum_{n=2}^{\infty} \Delta\bar{g}_n \int_{\phi=0}^{2\pi} \int_{\theta=0}^{\pi} [P_n(\cos\theta)]^2 \sin\theta\, d\theta\, d\phi$$

$$(3.117)$$

Let $\cos\theta = x$, then $-\sin\theta\, d\theta = dx$, and, on integrating with respect to ϕ, we have

$$\iint_S \Delta g(\theta,\phi) P_n(\cos\theta) d\Omega = 2\pi \sum_{n=2}^{\infty} \Delta\bar{g}_n \int_{x=-1}^{1} [P_n(x)]^2\, dx = 4\pi \frac{\Delta\bar{g}_n}{2n+1}$$

$$(3.118)$$

The last step uses the normalization of the Legendre polynomials (Section 1.13.2).

We can now obtain $\Delta\bar{g}_n$ from (3.118) and insert it into (3.114) to find the potential W of the geoid elevation. Using (3.101), we get the height of the geoid undulation:

$$h = \frac{R}{4\pi g} \sum_{n=2}^{\infty} \iint_S \frac{2n+1}{n-1} \left(\frac{R}{r}\right)^{n+1} P_n(\cos\theta) \Delta g(\theta,\phi) d\Omega \qquad (3.119)$$

The summation under the integration reduces to a function of the angle θ only, which we designate $F(\theta)$. With this function the height of the geoid is

$$h = \frac{R}{4\pi g} \iint_S F(\theta) \Delta g(\theta,\phi) dS \qquad (3.120)$$

This is known as Stokes' formula for the height of the geoid.

3.6.3 Evaluation of the function $F(\theta)$

The function $F(\theta)$ in Stokes' formula for the height of the geoid is the value, on the surface of the Earth, of the function $F(r,\theta)$ in the integrand of (3.119), given by

$$F(r, \theta) = \sum_{n=2}^{\infty} \frac{2n+1}{n-1} \left(\frac{R}{r}\right)^{n+1} P_n(\cos\theta) \qquad (3.121)$$

In order to simplify this expression we use the reciprocal-distance definition of the Legendre polynomials, in the alternative form developed in Box 1.4:

$$\frac{1}{u} = \frac{1}{r}\sum_{n=0}^{\infty} \left(\frac{R}{r}\right)^n P_n(\cos\theta) = \frac{1}{r} + \frac{R\cos\theta}{r^2} + \sum_{n=2}^{\infty} \frac{R^n}{r^{n+1}} P_n(\cos\theta) \qquad (3.122)$$

After altering the sequence, this allows us to write

$$\sum_{n=2}^{\infty} \frac{R^n}{r^{n+1}} P_n(\cos\theta) = \frac{1}{u} - \frac{1}{r} - \frac{R\cos\theta}{r^2} \qquad (3.123)$$

Expanding the sum in (3.121) gives

$$F(r, \theta) = 2\sum_{n=2}^{\infty} \left(\frac{R}{r}\right)^{n+1} P_n(\cos\theta) + 3\sum_{n=2}^{\infty} \frac{1}{n-1} \left(\frac{R}{r}\right)^{n+1} P_n(\cos\theta) \qquad (3.124)$$

The first term on the right is simply $2R$ times the left-hand side of (3.123).
To evaluate the second term on the right we note that

$$\frac{1}{r^2}\int_r^{\infty} \frac{dr}{r^n} = \frac{1}{n-1}\left(\frac{1}{r^{n+1}}\right) \qquad (3.125)$$

This relationship can be used to change the second expression on the right of (3.124) to

$$3\sum_{n=2}^{\infty} \frac{1}{n-1}\left(\frac{R}{r}\right)^{n+1} P_n(\cos\theta) = \frac{3}{r^2}\int_r^{\infty} \sum_{n=2}^{\infty} \frac{R^{n+1}}{r^n} P_n(\cos\theta) dr$$

$$= \frac{3R}{r^2}\int_r^{\infty} r\sum_{n=2}^{\infty} \frac{R^n}{r^{n+1}} P_n(\cos\theta) dr$$

Now we can substitute from (3.123):

$$3 \sum_{n=2}^{\infty} \frac{1}{n-1} \left(\frac{R}{r}\right)^{n+1} P_n(\cos\theta) = \frac{3R}{r^2} \int_r^{\infty} \left(\frac{r}{u} - 1 - \frac{R\cos\theta}{r}\right) dr$$

$$= \frac{3R}{r^2} \left\{ \int_r^{\infty} \frac{r\,dr}{u} - [r + R\cos\theta \log r]_r^{\infty} \right\}$$

(3.126)

The integration on the right must be done in several steps because the denominator u is a function of r. We must first rewrite the equation in a more tractable form:

$$\int_r^{\infty} \frac{r\,dr}{u} = \int_r^{\infty} \frac{r\,dr}{\sqrt{r^2 - 2rR\cos\theta + R^2}} = \int_r^{\infty} \frac{(r - R\cos\theta) + R\cos\theta}{\sqrt{(r - R\cos\theta)^2 + R^2\sin^2\theta}} dr$$

(3.127)

$$\int_r^{\infty} \frac{r\,dr}{u} = \int_r^{\infty} \frac{(r - R\cos\theta)\,dr}{\sqrt{(r - R\cos\theta)^2 + R^2\sin^2\theta}} + \int_r^{\infty} \frac{R\cos\theta\,dr}{\sqrt{(r - R\cos\theta)^2 + R^2\sin^2\theta}}$$

(3.128)

Next, we carry out each of these integrations separately: the first part is simply

$$\int \frac{(r - R\cos\theta)\,dr}{\sqrt{(r - R\cos\theta)^2 + R^2\sin^2\theta}} = \sqrt{(r - R\cos\theta)^2 + R^2\sin^2\theta} = u \quad (3.129)$$

For the second part we make use of the following standard integration:

$$\int \frac{a}{\sqrt{y^2 + b^2}} dy = a\log\left(y + \sqrt{y^2 + b^2}\right)$$

(3.130)

Letting $y = r - R\cos\theta$, $a = R\cos\theta$, and $b = R\sin\theta$ in this equation, the second integration becomes

$$\int \frac{R\cos\theta\,dr}{\sqrt{(r - R\cos\theta)^2 + R^2\sin^2\theta}} = R\cos\theta \cdot \log\left(r - R\cos\theta.\right.$$

$$\left. + \sqrt{r^2 - 2rR\cos\theta + R^2}\right)$$

$$= R\cos\theta \cdot \log(r - R\cos\theta + u) \quad (3.131)$$

Combining (3.128), (3.129), and (3.131) gives

$$\int\limits_{r}^{\infty} \frac{r\,dr}{u} = [u + R\cos\theta \cdot \log(r - R\cos\theta + u)]_{r}^{\infty} \tag{3.132}$$

Upon inserting this result into (3.126) we get

$$3\sum_{n=2}^{\infty} \frac{1}{n-1}\left(\frac{R}{r}\right)^{n+1} P_n(\cos\theta) = \frac{3R}{r^2}[u + R\cos\theta \cdot \log(r - R\cos\theta + u)$$
$$- r - R\cos\theta \cdot \log r]_{r}^{\infty} \tag{3.133}$$

At the limits of the integration we cannot insert $r = \infty$ directly. However, for very large r,

$$u = r\left(1 - \frac{2R\cos\theta}{r} + \frac{R^2}{r^2}\right)^{1/2} \approx r\left(1 - \frac{1}{2}\left(\frac{2R\cos\theta}{r}\right)\right) \approx r - R\cos\theta \tag{3.134}$$

Now we substitute this result into (3.133) to get the upper limit of the bracketed expression:

$$[u + R\cos\theta \cdot \log(r - R\cos\theta + u) - r - R\cos\theta \cdot \log r]^{\infty}$$
$$\approx -R\cos\theta + R\cos\theta \cdot \log(2(r - R\cos\theta)) - R\cos\theta \cdot \log r$$
$$\approx R\cos\theta \cdot \left(\log\left(2\frac{r - R\cos\theta}{r}\right) - 1\right)$$
$$\approx R\cos\theta \cdot (\log 2 - 1) \tag{3.135}$$

Evaluating both limits in (3.133) gives

$$3\sum_{n=2}^{\infty} \frac{1}{n-1}\left(\frac{R}{r}\right)^{n+1} P_n(\cos\theta)$$
$$= \frac{3R}{r^2}\left(R\cos\theta \cdot (\log 2 - 1) - u + r - R\cos\theta \cdot \log\left(\frac{r - R\cos\theta + u}{r}\right)\right) \tag{3.136}$$

Now we add this result to $2R$ times (3.123) to get the solution of (3.124):

$$F(r,\theta) = 2R\left(\frac{1}{u} - \frac{1}{r} - \frac{R\cos\theta}{r^2}\right)$$
$$+ \frac{3R}{r^2}\left(-R\cos\theta - u + r - R\cos\theta \cdot \log\left(\frac{r - R\cos\theta + u}{2r}\right)\right) \tag{3.137}$$

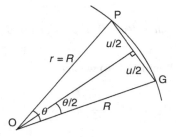

Fig. 3.7. Geometry for calculation of the geoid height at a point P from gravity measurements. G is a point on the surface of the Earth at which gravity was measured.

The point P at which the geoid height is to be calculated and the point G at which a gravity measurement is known lie on the surface of the Earth, where $r = R$, as in Fig. 3.7. These points form an isosceles triangle with the center of the Earth at O, so that $u = 2R \sin(\theta/2)$ and

$$\frac{r - R\cos\theta + u}{2r} = \frac{1}{2}\left(1 - \cos\theta + 2\sin\left(\frac{\theta}{2}\right)\right) = \sin\left(\frac{\theta}{2}\right) + \sin^2\left(\frac{\theta}{2}\right) \tag{3.138}$$

On substituting into (3.137), and noting that on the surface of the sphere $F(r,\theta)$ becomes $F(\theta)$, we have

$$F(\theta) = 2\left(\frac{1}{2\sin(\theta/2)} - 1 - \cos\theta\right)$$
$$+ 3\left(-\cos\theta - 2\sin\left(\frac{\theta}{2}\right) + 1 - \cos\theta \cdot \log\left(\sin\left(\frac{\theta}{2}\right) + \sin^2\left(\frac{\theta}{2}\right)\right)\right) \tag{3.139}$$

$$F(\theta) = \frac{1}{\sin(\theta/2)} + 1 - 6\sin\left(\frac{\theta}{2}\right) - 5\cos\theta$$
$$- 3\cos\theta \cdot \log\left(\sin\left(\frac{\theta}{2}\right) + \sin^2\left(\frac{\theta}{2}\right)\right) \tag{3.140}$$

The function $F(\theta)$ is plotted in Fig. 3.8. It has a singularity at $\theta = 0$, which must be excluded from the computation. $F(\theta)$ decreases rapidly with increasing angle θ for $\theta < 30°$ but still has an appreciable value at large angles, which means that distant gravity measurements can have an influence on the calculated geoid height.

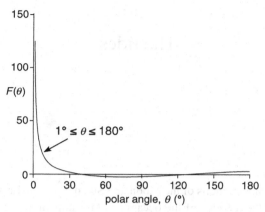

Fig. 3.8. Variation with angular distance θ of the function $F(\theta)$ in Stokes' formula for the height of the geoid.

FURTHER READING

Bullen, K. E. (1975). *The Earth's Density*. London: Chapman and Hall, 420 pp.
Groten, E. (1979). *Geodesy and the Earth's Gravity Field*. Bonn: Dümmler, 409 pp.
Hofmann-Wellenhof, B. and Moritz, H. (2006). *Physical Geodesy*, 2nd edn. Vienna: Springer, 403 pp.
Torge, W. (1989). *Gravimetry*. Berlin: de Gruyter, 465 pp.

4

The tides

The gravitational attractions of the Moon and Sun deform the Earth, giving rise to the periodic fluctuations of the oceanic surface known as the marine tides. The same forces also give rise to bodily tides in the solid Earth. The Moon's mass is much smaller than that of the Sun, but the lunar tidal effect is greater than the Sun's, because the Moon is much closer to the Earth. We first analyze the lunar tides, then take account of the solar tidal effects.

4.1 Origin of the lunar tide-raising forces

The lunar tidal forces arise from two sources: the gravitational attraction of the Moon on the Earth, and the joint rotation of the Earth and Moon about their common center of mass, which is called the *barycenter*. The barycenter moves around the Sun along Earth's orbit.

To find the location of the barycenter of the Earth–Moon system, let the distance between Earth and Moon be r_L, the mass of the Earth E, and the mass of the Moon M. If the barycenter B is at distance d from the center of the Earth, then, taking moments about B,

$$Ed = M(r_L - d) \tag{4.1}$$

and hence

$$d = \frac{M}{E + M} r_L \tag{4.2}$$

The mass-ratio of Moon and Earth M/E is equal to 0.0123, and the distance between Earth and Moon is 384,400 km, so the distance d is 4,670 km; i.e., the barycenter lies within the Earth. The center of the Earth moves around this point with the same rotational angular velocity ω_L as does the Moon (Fig. 4.1), and describes a circle with radius d.

116

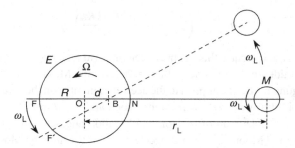

Fig. 4.1. Geometry of Earth and Moon in the plane of the Moon's orbit. The barycenter of the rotation is at B; ω_L is the rotation rate of the Moon about its axis and about the Earth; Ω is the Earth's own rotation rate, assumed normal to the Moon's orbit.

Let the Earth–Moon barycenter be at B and the center of the Earth at O; let the Earth's radius be R, the Moon's mass be M, and the distance between the centers of Earth and Moon be r_L, as in Fig. 4.1. At the center of the Earth, the gravitational acceleration a_O towards the Moon exactly balances the centrifugal acceleration $a_c = \omega_L^2 d$ of the Earth's motion around the circle with radius d, thus

$$\frac{GM}{r_L^2} = \omega_L^2 d \qquad (4.3)$$

The point F on the far side of the Earth is at distance $r_L + R$ from the Moon and $R + d$ from the barycenter. The gravitational acceleration at F towards the Moon is balanced by the centrifugal acceleration away from the Moon, and the net acceleration at F *towards* the Moon is

$$a_F = \frac{GM}{(r_L + R)^2} - \omega_L^2 (R + d) \qquad (4.4)$$

Applying the binomial expansion up to fourth order gives

$$a_F = \left(\frac{GM}{r_L^2} - 2\frac{GMR}{r_L^3} + 3\frac{GMR^2}{r_L^4} \right) - \omega_L^2 d - \omega_L^2 R \qquad (4.5)$$

The term $\omega_L^2 d$ is again the centrifugal acceleration of a rotation about a circle with radius d, and is directed away from the Moon. The centrifugal acceleration $\omega_L^2 R$ is also directed away from the Moon. It corresponds to motion of the point F about a circle with radius R. This rotation displaces F to F' in Fig. 4.1 and is a component of the Earth's rotation about its own axis. It does not contribute to the lunar tidal acceleration. Omitting this term and using the result of (4.3), we have for the tide-raising acceleration at F

$$a_F = -\left(2\frac{GMR}{r_L^3} - 3\frac{GMR^2}{r_L^4}\right) \tag{4.6}$$

The negative sign indicates that the net acceleration at F is *away* from the Moon. This causes a tide on the far side of the Earth from the Moon.

Similar arguments can be applied to the accelerations at N on the near side of the Earth, which is at distance $r_L - R$ from the Moon and $R - d$ from the barycenter. The centrifugal acceleration of the common rotation augments the gravitational acceleration of the Moon, and the net acceleration a_N towards the Moon is

$$a_N = \frac{GM}{(r_L - R)^2} + \omega_L^2(R - d) \tag{4.7}$$

The binomial expansion leads to the following equation for the acceleration at N towards the Moon:

$$a_N = \left(\frac{GM}{r_L^2} + 2\frac{GMR}{r_L^3} + 3\frac{GMR^2}{r_L^4}\right) - \omega_L^2 d + \omega_L^2 R \tag{4.8}$$

As before, the bracketed term is the lunar gravitational attraction and the centrifugal acceleration $\omega_L^2 d$ is away from the barycenter. The centrifugal acceleration $\omega_L^2 R$ is now directed *towards* the Moon, as expected for a rotation about the Earth's axis. The tide-raising acceleration at N is

$$a_N = \left(2\frac{GMR}{r_L^3} + 3\frac{GMR^2}{r_L^4}\right) \tag{4.9}$$

This acceleration acts towards the Moon and is responsible for the tide on the near side of the Earth.

The balance of the tidal forces is summarized in Fig. 4.2. The centrifugal acceleration $\omega_L^2 d$ away from the Moon is present at all points of the Earth.

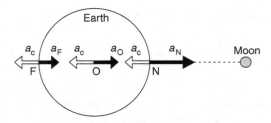

Fig. 4.2. Accelerations responsible for the lunar tides on the Earth: a_F, a_O, and a_N are the gravitational accelerations of the Moon at the furthest point (F), center of the Earth (O), and nearest point (N) to the Moon; a_c is the constant acceleration due to the Earth's rotation about the barycenter, excluding the component of this rotation about Earth's own axis.

It arises from the rigid-body rotation of the Earth about the barycenter (see Lowrie (2007) for a graphical explanation).

Comparison of (4.6) and (4.9) shows that the tidal accelerations at F and N are unequal. As a result, the lunar tide on the near side of the Earth is higher than that on the far side. A more detailed analysis of the tidal components and the direction of the tide-raising forces on the Earth is obtained by examining the tidal potential.

4.2 Tidal potential of the Moon

The calculation of the potential of the Moon's gravitational attraction at a point in the Earth (Fig. 4.3) is similar to the development of MacCullagh's formula. Spherical polar coordinates are centered at the center of the Earth. The lunar potential is calculated for a point P in the Earth at distance r from Earth's center. The radius to P makes an angle ψ with the direction to the Moon, and the geometry has rotational symmetry about this axis. The lunar potential W at P is inversely proportional to the distance u of P from the center of the Moon. The reciprocal-distance formula introduces the Legendre polynomials to describe the potential:

$$W = -G\frac{M}{u} = -G\frac{M}{r_{\text{L}}}\left(1 + \sum_{n=1}^{\infty}\left(\frac{r}{r_{\text{L}}}\right)^{n} P_{n}(\cos\psi)\right) \qquad (4.10)$$

Upon expanding the first few terms in the summation we get

$$W = -G\frac{M}{r_{\text{L}}} - G\frac{Mr\cos\psi}{r_{\text{L}}^{2}} - G\frac{Mr^{2}P_{2}(\cos\psi)}{r_{\text{L}}^{3}} - G\frac{Mr^{3}P_{3}(\cos\psi)}{r_{\text{L}}^{4}} - \cdots$$

$$(4.11)$$

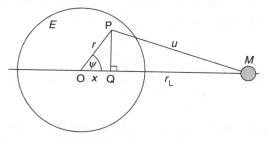

Fig. 4.3. Calculation of the lunar potential for a point P in the Earth at distance r from Earth's center and distance u from the Moon.

This equation is equivalent to a sum of individual potential terms, the first few of which give

$$W = W_0 + W_1 + W_2 + W_3 + \cdots \tag{4.12}$$

4.2.1 Significance of individual terms in the lunar potential

Potential W_0

$$W_0 = -G\frac{M}{r_L} \tag{4.13}$$

This first term in the sum is a constant, so its gradient is zero:

$$a_0 = -\nabla W_0 = 0 \tag{4.14}$$

This potential does not play a role in the tidal deformation of the Earth.

Potential W_1

$$W_1 = -G\frac{M(r\cos\psi)}{r_L^2} = -G\frac{M}{r_L^2}x \tag{4.15}$$

Here we have defined OQ in Fig. 4.3 as $x = r\cos\psi$. The x-axis is along the direction to the Moon. The gradient of the potential W_1 gives

$$a_1 = -\nabla W_1 = -\frac{\partial W_1}{\partial x} = \left(\frac{GM}{r_L^2}, 0, 0\right) \tag{4.16}$$

This acceleration acts in the direction of positive x, i.e., towards the Moon. It is independent of the position coordinates (r, ψ) and is therefore constant throughout the body of the Earth. It does not contribute to the tide-raising forces but balances the centrifugal acceleration of the Earth–Moon rotation about their common barycenter. An equal and opposite acceleration acts on the Moon and holds it in orbit around the Earth.

Potential W_2

$$W_2 = -G\frac{Mr^2 P_2(\cos\psi)}{r_L^3} \tag{4.17}$$

This is the potential of the main tidal deformation. It is much larger than all following terms and is regarded as the tidal potential, except in detailed analyses. It is proportional to the second-order Legendre polynomial

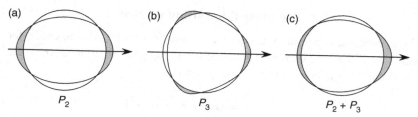

Fig. 4.4. Components of the lunar potential (not to scale): (a) main symmetric deformation proportional to a second-order Legendre polynomial; (b) next-largest component of deformation, proportional to a third-order Legendre polynomial; and (c) superposition of these components that gives rise to the diurnal tidal inequality.

$P_2(\cos \psi)$ and so has rotational symmetry about the Earth–Moon axis and gives equal tides on opposite sides of the Earth (Fig. 4.4(a)). For use in later discussions, let

$$A = -G\frac{M}{r_L^3} \tag{4.18}$$

This enables us to write the tidal potential in the more compact form

$$W_2 = Ar^2 P_2(\cos \psi) = Ar^2 P_2 \tag{4.19}$$

Potential W_3

$$W_3 = -G\frac{Mr^3 P_3(\cos \psi)}{r_L^4} \tag{4.20}$$

This potential describes a deformation with the symmetry of the third-order Legendre polynomial $P_3(\cos \psi)$. It is symmetric about the Earth–Moon axis but results in a tidal elevation on Earth's near side and a tidal depression on Earth's far side (Fig. 4.4(b)). Together with W_2 it describes the unequal diurnal tides explained in Section 4.1 (Fig. 4.4(c)). W_3 is the second-largest term in the tidal deformation, but is much smaller than W_2, as can be shown by forming the ratio of the two potentials:

$$\frac{W_2}{W_3} = \frac{r^2 P_2(\cos \psi)}{r_L^3}\frac{r_L^4}{r^3 P_3(\cos \psi)} = \left(\frac{r_L}{r}\right)\left(\frac{P_2}{P_3}\right) \geq 80 \tag{4.21}$$

This and higher-order terms in the tidal potential are usually disregarded except in detailed evaluation of the tidal heights.

4.2.2 The lunar tide-raising acceleration

The tide-raising acceleration is equal to the gradient of the tidal potential, for which we will use the dominant potential W_2. Using polar coordinates (r, ψ) the acceleration has a radial component a_r given by

$$a_r = -\frac{\partial W_2}{\partial r} = G\frac{M}{r_L^3}r(3\cos^2\psi - 1)$$

$$= G\frac{Mr}{2r_L^3} \cdot (1 + 3\cos(2\psi)) \tag{4.22}$$

The transverse component a_ψ is

$$a_\psi = -\frac{1}{r}\frac{\partial W_2}{\partial \psi} = G\frac{M}{r_L^3}r\frac{\partial}{\partial \psi}\frac{1}{2}(3\cos^2\psi - 1)$$

$$= -G\frac{Mr}{2r_L^3} \cdot 3\sin(2\psi) \tag{4.23}$$

These accelerations cause tidal displacements that are vertical (i.e., radial) on the Earth–Moon axis at $\psi = 0$ and $\psi = \pi$, as well as at an angular distance $\psi = \pm\pi/2$ from the axis. At intermediate locations the tide-raising forces have a horizontal as well as a radial component (Fig. 4.5).

4.2.3 The solar tide-raising acceleration

The tide-raising acceleration of the Sun can be described in a similar way to that of the Moon. The dependence of the lunar tidal amplitude on the Moon's mass

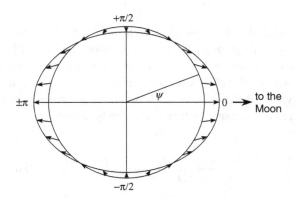

Fig. 4.5. Direction of the lunar tidal-raising force as a function of angular distance ψ from the Earth–Moon axis.

Table 4.1. *Rotational and orbital parameters of the Earth and Moon* (sources: *Groten, 2004; McCarthy and Petit, 2004).*

Parameter	Symbol	Units	Value
Mass of Sun	S	10^{30} kg	1.988 92
Heliocentric gravitational constant	GS	10^{14} m^3 s^{-2}	3.986 004 418
Mass of Earth	E	10^{24} kg	5.973 7
Geocentric gravitational constant	GE	10^{20} m^3 s^{-2}	1.327 124 4
Solar mass ratio, S/E	μ_S	10^5	3.329 46
Mass of Moon	M	10^{22} kg	7.347 7
Selenocentric gravitational constant	GM	10^{12} m^3 s^{-2}	4.902 799
Lunar mass ratio, M/E	μ_L		0.012 300 034
Mean geocentric radius of the Moon's orbit	r_L	10^8 m	3.844
Mean heliocentric radius of Earth's orbit	r_S	10^{11} m	1.495 874 4
Present rotation rate of the Earth	Ω_0	10^{-5} rad s^{-1}	7.292 1
Moment of inertia of Earth about its rotation axis	C	10^{37} kg m^2	8.019
Angular momentum of Earth–Moon system	h	10^{34} kg m^2 s^{-1}	3.435
Earth's mean radius	R	10^6 m	6.371 000 4
Moon's mean radius	R_L	10^6 m	1.738

and distance from the Earth is contained in the factor A defined in (4.18), which we will call A_L for this comparison. The tidal effect of the Sun depends on a similar factor A_S, in which the mass S of the Sun replaces the lunar mass M, and the Earth–Sun separation r_S replaces the Earth–Moon separation r_L. At any given point (r, ψ) on the Earth the ratio A_L/A_S expresses the relative effects of the lunar and solar tide-raising accelerations:

$$\frac{a_L}{a_S} = \frac{A_L}{A_S} = \frac{-GM/r_L^3}{-GS/r_S^3} = \frac{M}{S}\left(\frac{r_S}{r_L}\right)^3 = 2.2 \qquad (4.24)$$

The masses of Sun and Moon, and their distances from the Earth are listed in Table 4.1. The ratio of the Sun's mass to the Moon's mass (S/M) is about 27,000,000. The ratio of the Sun's distance to the Moon's distance (r_S/r_L) is 389. However, in comparing the lunar and solar tidal effects the distance-ratio is cubed, which attenuates the tidal effect of the Sun more than it does that of the Moon. Consequently, the Sun is responsible for only about one third of the observed tide, with two thirds being caused by the Moon.

The lunar and solar tidal accelerations depend on the relative phases of the Sun and Moon. When they are aligned, on the same side of the Earth (known as *conjunction*) or on opposite sides (*opposition*), their tidal accelerations reinforce each other and give rise to extra-high *spring tides*. When the directions to Sun and Moon are perpendicular, the tidal accelerations are in *quadrature* and tend to cancel each other out partially, causing extra-low *neap tides*.

4.3 Love's numbers and the tidal deformation

When we think of the tides, we usually mean the observed semidiurnal rise and fall of the ocean surface. The marine tide is an elastic response of the Earth as a whole to the lunar deforming potential. However, the tide is measured with respect to the solid Earth, which is also deformed by the lunar gravitation. The observed tide is the difference. The marine and bodily tides are characterized by global elastic constants called Love's numbers.

4.3.1 Tidal height

Let the elevation of the equipotential surface due to W_2 at any particular point be H_0. The uplift takes place against the acceleration of gravity, so the work done (gH_0) is equal to the change in potential. The height of the elevation is given by

$$H_0 = \frac{W_2}{g} \tag{4.25}$$

Tidal deformations are the elastic response of the Earth to the lunar deforming forces. The redistribution of mass gives rise to an additional potential, which must be taken into account in analyzing the tidal potential. In 1911, A. E. H. Love, an English mathematician, reasoned that the extra potential U_2 of the deformation should be proportional to the deforming potential W_2, i.e.,

$$U_2 = kW_2 \tag{4.26}$$

The proportionality constant k is a global value for the elastic response of the Earth as a whole. The added potential enhances the total tidal potential to $(1 + k)W_2$ and increases the vertical tidal displacement to H_1 (Fig. 4.6):

$$H_1 = \frac{W_2 + U_2}{g} = (1 + k)\frac{W_2}{g} \tag{4.27}$$

The solid body of the Earth is involved in the tidal response. The potential of the solid surface displacement is also proportional to the perturbing potential

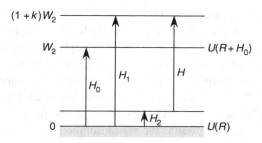

Fig. 4.6. Factors involved in computation of the height of the equilibrium tide on an elastic Earth. W_2 is the lunar tidal potential and k is Love's first number.

W_2, with proportionality constant h, so the height H_2 of the bodily tide can be expressed as

$$H_2 = h \frac{W_2}{g} \tag{4.28}$$

On combining the results, the height H of the equilibrium tide is seen with reference to Fig. 4.6 to be

$$H = H_1 - H_2 = (1 + k - h) \frac{W_2}{g} = \alpha H_0 \tag{4.29}$$

where

$$\alpha = 1 + k - h \tag{4.30}$$

Here α is the ratio of the observed vertical tidal height to the theoretical height on a rigid Earth ($k = h = 0$). Empirical values can be obtained from direct measurements of tidal height. However, restrictive conditions for direct tidal observations must be observed. The body of water must be small enough that it has a short reaction time to the perturbing potential and there is no phase lag. The shape and bathymetry of the body of water must not amplify the tidal effects. For these reasons enclosed bodies of water with natural periods less than a day have been favored in direct measurements. These give a value $\alpha \approx 0.7$.

4.3.2 Tidal gravity anomaly

The lunar tidal attraction affects measurements of gravity made on Earth, necessitating a tidal correction. The tidal gravity anomaly derives from three potentials that affect a gravimeter set up on the Earth's surface: (1) the geopotential, (2) the lunar tidal potential, and (3) the potential of the tidal deformation. For the first of these it is adequate to substitute the Earth's gravitational potential, while the second potential is the lunar deforming potential W_2. As explained in the previous section, the lunar tide corresponds to a mass redistribution within the Earth, which has a potential kW_2. We need to determine the potential of this deformation outside the Earth on the measurement surface.

Equation (4.19) shows that the deformation potential kW_2 is equal to $kAr^2 P_2(\cos \psi)$. This is a solution of Laplace's equation for a space in which r can be zero, i.e., inside the Earth. We seek a solution that is valid outside the Earth. In general, a potential Φ satisfying Laplace's equation may be written

$$\Phi = \left(Ar^2 + \frac{B}{r^3} \right) P_2(\cos \psi) \tag{4.31}$$

We separate this potential Φ into two potentials for different realms:

$$\begin{aligned} \Phi_i &= Ar^2 P_2(\cos \psi), \quad r < R \\ \Phi_e &= \frac{B}{r^3} P_2(\cos \psi), \quad r \geq R \end{aligned} \tag{4.32}$$

The first part, Φ_i, is valid inside the Earth, where r can be zero; the second part, Φ_e, is valid outside the Earth, where r can be infinite. The two solutions vary differently with radial distance. At the same azimuth ψ from the symmetry axis they are in the ratio

$$\frac{\Phi_e}{\Phi_i} = \frac{B/r^3}{Ar^2} = \left(\frac{B}{A} \right) \frac{1}{r^5} \tag{4.33}$$

The potential must be continuous at the Earth's surface, i.e., $\Phi_e = \Phi_i$ where $r = R$, thus

$$\frac{B}{A} = R^5 \tag{4.34}$$

and

$$\Phi_e = \left(\frac{R}{r} \right)^5 \Phi_i \tag{4.35}$$

By applying this result to the lunar tidal deformation, we find that its potential inside the Earth is kW_2, so its potential outside the Earth is $kW_2(R/r)^5$. Thus the potential U_T of the tidal gravity anomaly, as measured outside the Earth, is

$$U_T = -G\frac{E}{r} + W_2 + kW_2 \left(\frac{R}{r} \right)^5 \tag{4.36}$$

The first term represents the gravity potential of the undeformed Earth, the second term that of the Moon. The third term is the gravity potential associated with the tidal deformation. The acceleration due to gravity is the radial gradient due to U_T:

$$g(r) = -\frac{\partial U_T}{\partial r} = -G\frac{E}{r^2} - \frac{\partial}{\partial r}W_2 - k\frac{\partial}{\partial r}W_2\left(\frac{R}{r}\right)^5 \qquad (4.37)$$

Each term must be evaluated at the surface of the solid Earth. The tidal displacement of the solid surface (4.28) raises this to the position

$$r = R + H_2 = R\left(1 + h\frac{H_0}{R}\right) \qquad (4.38)$$

The tidal elevation H_0 is very small compared with the Earth's radius, so we can make use of the binomial expansion to first order, by writing

$$\left(1 + h\frac{H_0}{R}\right)^n \approx 1 + nh\frac{H_0}{R} \qquad (4.39)$$

On differentiating the first term in (4.36) and using this simplification, we get

$$-G\frac{E}{r^2}\bigg|_{r=R(1+hH_0/R)} = -G\frac{E}{R^2}\left(1 + h\frac{H_0}{R}\right)^{-2}$$

$$\approx g(R)\left(1 - 2h\frac{H_0}{R}\right) \qquad (4.40)$$

Differentiating the second term and neglecting terms of order $(H_0/R)^2$ and higher gives

$$-\frac{\partial}{\partial r}W_2 = -\frac{\partial}{\partial r}Ar^2 P_2(\cos\psi) = -2\frac{W_2}{r}\bigg|_{r=R(1+hH_0/R)}$$

$$= -2g\frac{H_0}{R}\left(1 - h\frac{H_0}{R}\right)$$

$$\approx -2g(R)\frac{H_0}{R} \qquad (4.41)$$

By applying the same rules to expand the third term in (4.37) we obtain

$$-k\frac{\partial}{\partial r}W_2\left(\frac{R}{r}\right)^5 = -kAP_2(\cos\psi)\frac{\partial}{\partial r}\frac{R^5}{r^3}$$

$$= 3kAP_2(\cos\psi)\frac{R^5}{r^4} \qquad (4.42)$$

$$3kAP_2(\cos\psi)\frac{R^5}{r^4}\Big|_{r=R(1+hH_0/R)} = 3kAP_2(\cos\psi)\frac{R^5}{R^4}\left(1-4h\frac{H_0}{R}\right)$$

$$\approx 3k\frac{W_2}{R}\left(1-4h\frac{H_0}{R}\right) \tag{4.43}$$

$$-k\frac{\partial}{\partial r}W_2\left(\frac{R}{r}\right)^5 \approx 3kg(R)\frac{H_0}{R} \tag{4.44}$$

On combining the results of (4.40), (4.41), and (4.44) we have

$$g(r) = g(R)\left(1-2h\frac{H_0}{R}\right) - 2g(R)\frac{H_0}{R} + 3kg(R)\frac{H_0}{R} \tag{4.45}$$

$$g(r) = g(R)\left(1-2\frac{H_0}{R}-2h\frac{H_0}{R}+3k\frac{H_0}{R}\right) \tag{4.46}$$

The difference between $g(r)$ and $g(R)$ is the gravity anomaly Δg caused by the lunar tide on the deformed Earth:

$$\Delta g = g(r) - g(R) = -2g(R)\frac{H_0}{R}\left(1+h-\frac{3}{2}k\right) \tag{4.47}$$

If the Earth were rigid ($k = h = 0$) and unable to deform in response to the lunar tidal forces, there would still be a tidal gravity anomaly, corresponding to the gravitational attraction of the Moon

$$\Delta g_0 = -2g(R)\frac{H_0}{R} \tag{4.48}$$

Thus,

$$\Delta g = \Delta g_0\left(1+h-\frac{3}{2}k\right) = \beta\Delta g_0 \tag{4.49}$$

where

$$\beta = 1+h-\frac{3}{2}k \tag{4.50}$$

is the ratio of the observed tidal gravity anomaly on the deformed Earth to the theoretical value for a rigid Earth. Direct measurements give $\beta \approx 1.15$.

The simultaneous solution of (4.30) and (4.50) using the measured values for α and β yields values $k \approx 0.3$ and $h \approx 0.6$ for the Love numbers.

4.3.3 Tidal deflection of the vertical

The horizontal component of the tide-raising acceleration (Fig. 4.5) produces a horizontal tidal displacement. As before, the tidal potential W_2 is enhanced by the tidal bulge to $(1 + k)W_2$. In 1912 T. Shida introduced the number l to account for the potential of the horizontal tide, which, analogously to Love's number h, is proportional to the deforming potential W_2. The complete potential of the horizontal tide is then

$$W_h = (1 + k - l)W_2 \qquad (4.51)$$

The effect of the horizontal tide is to deflect the vertical direction. The deforming tidal potential W_2 produces horizontal components of gravity g_ψ and g_ϕ in the directions of increasing polar angle ψ and longitude ϕ, respectively. At the Earth's surface $r = R$ these are given by

$$g_\psi = -\frac{1}{R}\frac{\partial W_h}{\partial \psi}$$
$$g_\phi = -\frac{1}{R \sin \psi}\frac{\partial W_h}{\partial \phi} \qquad (4.52)$$

The vertical direction is deflected by amounts φ_ψ and φ_ϕ corresponding to the angles formed between the horizontal components of gravity and the radial component:

$$\varphi_\psi \approx \tan \varphi_\psi = \frac{g_\psi}{g}$$
$$\varphi_\phi \approx \tan \varphi_\phi = \frac{g_\phi}{g} \qquad (4.53)$$

The deflections of the vertical of tidal origin are obtained by combining (4.51), (4.52), and (4.53):

$$\varphi_\psi = -(1 + k - l)\frac{1}{gR}\frac{\partial W_2}{\partial \psi}$$
$$\varphi_\phi = -(1 + k - l)\frac{1}{gR \sin \psi}\frac{\partial W_2}{\partial \phi} \qquad (4.54)$$

On a rigid Earth $k = l = 0$ and the deflections of the vertical are

$$(\varphi_\psi)_0 = -\frac{1}{gR}\frac{\partial W_2}{\partial \psi}$$
$$(\varphi_\phi)_0 = -\frac{1}{gR \sin \psi}\frac{\partial W_2}{\partial \phi} \qquad (4.55)$$

The quantity

$$\chi = 1 + k - l \tag{4.56}$$

represents the ratio of the observed deflection of the vertical caused by the lunar tide on an elastic Earth to the theoretical deflection for a rigid Earth. Analysis of the tidal deflection of the vertical shows that Shida's number is a very small quantity ($l \approx 0.08$).

4.3.4 Satellite-derived values for k, h, and l

Satellite observations have replaced direct measurement as a means of determining the Love and Shida numbers. The tidal deformations of the geopotential cause slight perturbations of satellite orbits. The observed satellite orbits are compared with what would be expected for a model Earth. The models have to incorporate some assumptions, namely that the Earth is spherical, non-rotating, elastic, and isotropic. The elastic constants then vary only with depth, and may be interpreted from observations of seismic travel times. The most widely used is the Preliminary Reference Earth Model (PREM) (Dziewonski and Anderson, 1981). The satellite-derived values of Love's number and Shida's number for the ellipsoidal tidal deformation are $k = 0.2980$, $h = 0.6032$, and $l = 0.0839$.

4.4 Tidal friction and deceleration of terrestrial and lunar rotations

The tidal bulge of the Earth is, to a first approximation, a prolate ellipsoid with symmetry axis aligned with the Earth–Moon axis. This configuration would give high tides at positions directly under the Moon and on the opposite side of the Earth. However, for several reasons the reaction of the Earth to the tidal forces is delayed. This is partly because the response of the solid Earth to forces on the timescale of the tides is not perfectly elastic. Also, the redistribution of water in the oceans is hindered by its viscosity, as well as by the presence of islands, bays, and uneven bottom topography. These interactions act as a frictional resistance that delays the tidal deformation. During the delay time the Earth's own rotation carries the tidal bulge forward. By the time the bulge has reached its peak height the axis of the tidal bulge has advanced about 2.9° past the Earth–Moon axis (Fig. 4.7).

Suppose the excess mass in the tidal bulge at Q to be concentrated at a point. The gravitational attraction of the Moon exerts a force F_2 on this part of the bulge. Similarly, a force F_1 acts on the part of the bulge at P. Because Q is closer

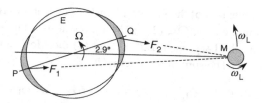

Fig. 4.7. Relationship of the torque that decelerates the Earth's rotation to the delay of the lunar tidal bulge due to inelastic and frictional effects.

to the Moon than P, the force F_2 is stronger than F_1; also, the acute angle at Q is larger than the acute angle at P, so the component of F_2 normal to the axis of the tidal bulge is larger than that of F_1. The forces cause a torque on the spinning Earth opposite to its direction of rotation. The frictional torque slows the Earth's rotation, causing the length of the day to increase by about 2.4 seconds per century. To maintain constant angular momentum of the closed Earth–Moon system, the rates of rotation of the Moon about its axis and about the Earth also decrease, and the Earth–Moon separation increases. The Moon's rotation rate about its axis has decreased to the extent that it is now synchronous with its rotation rate about the Earth. As a result an observer on Earth always seems to see the same face of the Moon.

In fact, the maximum amount of the Moon's surface visible at any time from the Earth is about 40%, because the curvature of the Moon's surface means that the periphery of the lunar globe is not visible from Earth. However, the Moon's orbit is slightly elliptical, its axis is slightly tilted to the pole to its orbit around the Earth, and due to Earth's rotation an observer views the Moon from slightly different angles at different times of day. These effects cause irregularities in the Moon's motion as viewed from Earth – called *librations* – that over time enable us to see 59% of the Moon's surface.

4.4.1 Angular momentum of the Earth–Moon system

The dimensions and rates of rotation of the Earth and the Moon, their separation, and the location of their barycenter are shown schematically in Fig. 4.1, as viewed from above the orbital plane of the Moon; the values of these parameters are given in Table 4.1. The focus of the orbit is at the barycenter, which is at a distance d from the mid-point of the Earth and at $r_L - d$ from the mid-point of the Moon. Let the moment of inertia of the Earth about its rotation axis be C and that of the Moon about its axis be C_L. The rotation axes are assumed to be perpendicular to the orbital plane.

The angular momentum of the system consists of contributions from (1) the Earth about its rotation axis, $C\Omega$; (2) the Moon about its rotation axis, $C_L\omega_L$; (3) the Earth about the barycenter, $Ed^2\omega_L$; and (4) the Moon about the barycenter, $M(r_L - d)^2\omega_L$. The sum of these terms is

$$h = C\Omega + C_L\omega_L + Ed^2\omega_L + M(r_L - d)^2\omega_L \qquad (4.57)$$

It was shown in Section 3.3.2 that the moment of inertia of a sphere is proportional to its mass times the square of its radius. The proportionality constants for most Earth-like planets are around 0.3, so the ratio of the angular momenta of the Earth and Moon can be estimated:

$$\frac{C_L\omega_L}{C\Omega} \approx \frac{M}{E}\left(\frac{R_L}{R}\right)^2 \frac{\omega_L}{\Omega} \approx \frac{1}{81} \cdot \frac{1}{13} \cdot \frac{1}{27} \approx 3.3 \times 10^{-5} \qquad (4.58)$$

In this comparison the lunar mass ratio is $M/E = 0.0123 = 1/81$, the equatorial radius of the Moon is $R_L \approx 1{,}738$ km, that of the Earth is $R = 6{,}378$ km, and the lunar sidereal rotation rate is 27.3 days. The very small value of the ratio shows that the angular momentum of the Moon's own rotation can be ignored in this discussion.

From (4.2) the distance of the center of the Moon from the barycenter is

$$r_L - d = \frac{E}{E + M}r_L \qquad (4.59)$$

By inserting this and (4.2) into (4.57), we get the angular momentum of the Earth–Moon system:

$$h = C\Omega + E\omega_L r_L^2\left(\frac{M}{E + M}\right)^2 + M\omega_L r_L^2\left(\frac{E}{E + M}\right)^2 \qquad (4.60)$$

$$h - C\Omega = \omega_L r_L^2\left(\frac{EM}{E + M}\right) \qquad (4.61)$$

4.4.2 Slowing of terrestrial and lunar rotations

Equation (4.61) has implications for the rates of rotation of the Earth and Moon. The gravitational attraction of the Earth on the Moon exactly balances the centrifugal acceleration of the Moon's orbital acceleration about the barycenter. This provides the additional equation

$$\frac{GE}{r_L^2} = \omega_L^2(r_L - d) \qquad (4.62)$$

and, on substituting for $(r_L - d)$ from (4.59), this becomes

$$\frac{GE}{r_L^2} = \omega_L^2 r_L \left(\frac{E}{E+M}\right) \qquad (4.63)$$

and thus

$$G(E+M) = \omega_L^2 r_L^3 \qquad (4.64)$$

This is, in fact, Kepler's Third Law for the Earth–Moon system. Now we square both sides, getting

$$G^2(E+M)^2 = \omega_L^4 r_L^6 \qquad (4.65)$$

Next we form the cube of (4.61),

$$(h - C\Omega)^3 = \omega_L^3 r_L^6 \left(\frac{EM}{E+M}\right)^3 \qquad (4.66)$$

Comparing (4.65) and (4.66) gives

$$(h - C\Omega)^3 = \frac{G^2(E+M)^2}{\omega_L} \frac{E^3 M^3}{(E+M)^3} \qquad (4.67)$$

Simplifying so that only the constant terms G, E, and M are on the right of the equation, we have

$$\omega_L (h - C\Omega)^3 = \frac{G^2 E^3 M^3}{E+M} \qquad (4.68)$$

The lunar tidal friction acts as a brake on the Earth's rotation, slowing it down and increasing the length of the day by about 2.4 ms per century. The total angular momentum of the system, h, is constant, as is the right-hand side of the equation. Thus, if Ω on the left-hand side of the equation is decreasing, the lunar rotation ω_L must also be decreasing. At the same time, in order to maintain (4.64), the distance between the Earth and Moon, r_L, must be increasing. At present the increase amounts to about 3.7 cm per year.

4.4.3 Development of the Earth–Moon separation

The tidal friction exerted by the Earth on the Moon has slowed the Moon's rotation until it is now synchronous with its orbital rotation around the Earth. Eventually the lunar tidal friction will slow the Earth's rotation so that it is also synchronous with the Moon's rotation. At that stage a terrestrial day, a lunar day, and the month will all have the same length. Meanwhile the Moon will continue to move further from the Earth. How far will the Moon be from the Earth when the rotations are synchronous? We can answer this question by setting $\omega_L = \Omega$ in (4.68). For convenience we also normalize the rotation in terms of Ω_0, the present rate of rotation of the Earth:

$$\frac{\Omega}{\Omega_0}\left(\frac{h}{C\Omega_0} - \frac{\Omega}{\Omega_0}\right)^3 = \frac{G^2 E^3 M^3}{C^3 \Omega_0^4 (E + M)} \tag{4.69}$$

Let the normalized rotation rate be $n = \Omega/\Omega_0$ and the normalized angular momentum be $a = h/(C\Omega_0)$, and let the expression on the right-hand side of the equation be b. Both a and b are constants, so we have to solve an equation with the form

$$n(a - n)^3 = b \tag{4.70}$$

This fourth-order equation in n has four roots, of which two are imaginary and of no interest, and two are real. The real roots, obtained numerically or graphically as in Box 4.1, are $n = 0.213$ and $n = 4.92$. The first solution

---

Box 4.1. Synchronous rotation of Earth and Moon

Equation (4.67) for the synchronous rotation of the Earth about its axis, the Moon about the Earth, and the Moon about its own axis can be written as

$$n(a - n)^3 = b \tag{1}$$

in which the normalized rotation rate is $n = \Omega/\Omega_0$, and the constants a and b are

$$a = \frac{h}{C\Omega_0} \tag{2}$$

$$b = \frac{G^2 E^3 M^3}{C^3 \Omega_0^4 (E + M)} \tag{3}$$

The numerical values of a and b are found by inserting the currently accepted values of the relevant parameters (Table 4.1) into the defining equation. This yields $a = 5.8742$ and $b = 4.272$. The equation becomes

$$n(5.8742 - n)^3 = 4.272 \tag{4}$$

The real roots of this fourth-order equation can be found by evaluating numerically the functions

$$F_1(n) = (5.8742 - n)^3 \tag{5}$$

$$F_2(n) = \frac{4.272}{n} \tag{6}$$

and finding the values of n that give $F_1(n) = F_2(n)$. Alternatively, the functions can be plotted as in Fig. B4.1 and the points of intersection of the curves determined.

The equation has only two real roots, which are $n = 0.0213$ and $n = 4.92$.

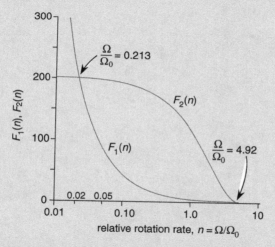

Fig. B4.1. Graphical solution for Ω, the synchronous rotation rate of the Earth and Moon; Ω_0 is the present rotation rate of the Earth.

corresponds to a rotation period of 47 days and an Earth–Moon separation of 87 times Earth's radius ($r_L = 87R$). The present distance between the centers of the Earth and Moon is 60 times Earth's radius, so this solution gives the conditions for a future synchroneity of the rotations. The second root gives a rotation period of 4.9 hr and a lunar distance of 2.3 times Earth's radius ($r_L = 2.3R$), corresponding to an earlier time in the Moon's history. However, this solution is unrealistic because it places the Moon within the Roche limit of the Earth, at which position the Earth's gravity would tear the Moon apart.

FURTHER READING

Lambeck, K. (1988). *Geophysical Geodesy: The Slow Deformations of the Earth.* Oxford: Clarendon Press, 718 pp.

Lowrie, W. (2007). *Fundamentals of Geophysics*, 2nd edn. Cambridge: Cambridge University Press, 381 pp.

Melchior, P. (1966). *The Earth Tides*. Oxford: Pergamon Press, 458 pp.

5

Earth's rotation

The Earth is not rigid and its rotation causes it to deform, flattening at the poles and bulging at the equator. The gravitational attractions of Sun and Moon on the equatorial bulge result in torques on the Earth, which cause additional motions of the rotation axis, known as *precession* and *nutation*. These motions occur relative to a coordinate system fixed in space, for example in the solar system. The rotation axis is inclined to the pole to the ecliptic plane at a mean angle of 23.425°; this angle is the *obliquity* of the axis. Precession is a very slow motion of the tilted rotation axis around the pole to the ecliptic, with a period of 25,720 yr. The nutation is superposed on this motion and consists of slight fluctuations in the rate of precession as well as in the obliquity.

The other planets also affect the Earth's rotation, causing small but significant cyclical changes on a very long timescale. These are observable directly by precise measurement of the position of the rotation axis using very-long-baseline interferometry (VLBI). The fluctuations influence the intensity of solar radiation incident on the Earth and produce cyclical climatic effects that are evident in sedimentary processes, where they are known as the Milankovitch (or Milanković) cycles. They correspond to retrograde precession of the rotation axis (period ~ 26 kyr), changes in the angle of obliquity (period ~ 41 kyr), prograde precession of Earth's elliptical orbit (period ~ 100 kyr), and variation of the ellipticity of the orbit (period ~ 100 kyr).

In addition to these phenomena, the Earth's rotation is affected on a shorter timescale by the planet's mass distribution. When the instantaneous rotation axis deviates from the axis of figure determined by the long-term rotation, a cyclical motion of the rotation axis about its mean position arises. This is known as the Chandler wobble. In contrast to the precession and nutation resulting from external forces, the wobble results from the imbalance in mass distribution with respect to the instantaneous rotation axis. It takes place in the Earth's coordinate system and is evident as small variations in latitude with a period of 435 days.

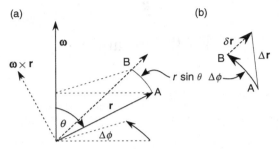

Fig. 5.1. Rotation of a displacement vector **r** inclined at angle θ to the rotation axis.

5.1 Motion in a rotating coordinate system

The displacement of a body on the rotating Earth may be considered to have two parts. The first is a simple displacement relative to coordinate axes defined for the Earth. The second arises from the rotation of the Earth relative to a fixed set of axes; these might be defined, for example, relative to the solar system.

5.1.1 Velocity

Consider an orthogonal spherical coordinate system with unit vectors (e_r, e_θ, e_ϕ). Let **r** be a displacement vector that makes an angle θ with the axis of rotation (Fig. 5.1(a)). If the Earth rotates about this axis with angular velocity **ω** relative to fixed axes, then, in an infinitesimal time Δt, the vector **r** rotates through an angle Δ_ϕ. This produces a rotational displacement $\Delta \mathbf{r}_1 = (r \sin \theta \, \Delta \phi) \, \mathbf{e}_\phi$ (Fig. 5.1(b)). If, in the same time, **r** undergoes a local incremental change $\delta \mathbf{r}$, the total displacement relative to the fixed coordinate system is

$$\Delta \mathbf{r} = \delta \mathbf{r} + \Delta \mathbf{r}_1 = \delta \mathbf{r} + (r \sin \theta \cdot \Delta \phi) \mathbf{e}_\phi \tag{5.1}$$

Dividing throughout by the time increment Δt gives the relationship between a velocity relative to the fixed axes and the velocity in the rotating system:

$$\frac{\Delta \mathbf{r}}{\Delta t} = \frac{\delta \mathbf{r}}{\Delta t} + \left(r \sin \theta \cdot \frac{\Delta \phi}{\Delta t} \right) \mathbf{e}_\phi \tag{5.2}$$

$$\frac{d \mathbf{r}}{dt} = \lim_{\Delta t \to 0} \left(\frac{\Delta \mathbf{r}}{\Delta t} \right) = \frac{\partial \mathbf{r}}{\partial t} + (r \sin \theta \cdot \omega) \mathbf{e}_\phi \tag{5.3}$$

The last term in (5.3) is equal to (**ω** × **r**), thus

$$\frac{d\mathbf{r}}{dt} = \frac{\partial \mathbf{r}}{\partial t} + (\boldsymbol{\omega} \times \mathbf{r}) \tag{5.4}$$

Thus, we have

$$\mathbf{v}_f = \mathbf{v} + (\boldsymbol{\omega} \times \mathbf{r}) \tag{5.5}$$

where $\mathbf{v}_f$ is the velocity relative to the fixed axes, $\mathbf{v}$ is the velocity in the rotating system, and $(\boldsymbol{\omega} \times \mathbf{r})$ is an additional velocity component due to the rotation of the moving set of axes.

5.1.2 Acceleration

Equation (5.4) can be rewritten as

$$\frac{d}{dt}\mathbf{r} = \left(\frac{\partial}{\partial t} + \boldsymbol{\omega} \times\right)\mathbf{r} \tag{5.6}$$

The expression in parentheses may be regarded as an operator acting on the vector $\mathbf{r}$. This allows us to express the acceleration as

$$\frac{d^2}{dt^2}\mathbf{r} = \frac{d}{dt}\left(\frac{d\mathbf{r}}{dt}\right) = \left(\frac{\partial}{\partial t} + \boldsymbol{\omega} \times\right)\left(\frac{\partial \mathbf{r}}{\partial t} + \boldsymbol{\omega} \times \mathbf{r}\right) \tag{5.7}$$

Evaluating the right-hand side step-by-step gives

$$\frac{d^2\mathbf{r}}{dt^2} = \frac{\partial^2\mathbf{r}}{\partial t^2} + \frac{\partial}{\partial t}(\boldsymbol{\omega} \times \mathbf{r}) + \left(\boldsymbol{\omega} \times \frac{\partial \mathbf{r}}{\partial t}\right) + (\boldsymbol{\omega} \times \boldsymbol{\omega} \times \mathbf{r}) \tag{5.8}$$

If we assume that the angular velocity $\boldsymbol{\omega}$ of the rotating system is constant, then

$$\frac{d^2\mathbf{r}}{dt^2} = \frac{\partial^2\mathbf{r}}{\partial t^2} + 2\left(\boldsymbol{\omega} \times \frac{\partial \mathbf{r}}{\partial t}\right) + (\boldsymbol{\omega} \times \boldsymbol{\omega} \times \mathbf{r}) \tag{5.9}$$

On rearranging terms, we get

$$\frac{\partial^2\mathbf{r}}{\partial t^2} = \frac{d^2\mathbf{r}}{dt^2} - (\boldsymbol{\omega} \times \boldsymbol{\omega} \times \mathbf{r}) - 2(\boldsymbol{\omega} \times \mathbf{v}) \tag{5.10}$$

or

$$\mathbf{a}_r = \mathbf{a}_f + \mathbf{a}_R + \mathbf{a}_C \tag{5.11}$$

where $\mathbf{a}_r = \partial^2\mathbf{r}/\partial t^2$ is the acceleration experienced by a moving object in the rotating system, and $\mathbf{a}_f = d^2\mathbf{r}/dt^2$ is the acceleration in the fixed coordinate system. The second acceleration on the right-hand side is $\mathbf{a}_R = -(\boldsymbol{\omega} \times \mathbf{r} \times \mathbf{r})$.

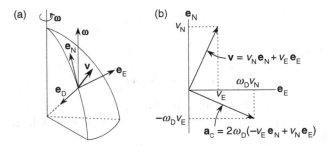

Fig. 5.2. (a) Directions of the north ($\mathbf{e}_N$), east ($\mathbf{e}_E$), and vertically downward ($\mathbf{e}_D$) unit vectors of orthogonal reference axes, and the horizontal velocity $\mathbf{v}$, in relation to the rotation vector $\boldsymbol{\omega}$. (b) Vectors in the horizontal plane, showing that the Coriolis acceleration $\mathbf{a}_C$ acts perpendicularly to the right of the direction of motion $\mathbf{v}$ in the northern hemisphere.

Inspection of the direction and magnitude of $\mathbf{a}_R$ shows that it is the familiar centrifugal acceleration. The final acceleration is

$$\mathbf{a}_C = -2(\boldsymbol{\omega} \times \mathbf{v}) = 2(\mathbf{v} \times \boldsymbol{\omega}) \tag{5.12}$$

$\mathbf{a}_C$ is called the Coriolis acceleration; it has important consequences for moving objects in a rotating framework.

5.2 The Coriolis and Eötvös effects

Suppose that a body is moving with horizontal velocity $\mathbf{v}$ on the surface of the Earth, which is rotating with angular velocity $\boldsymbol{\omega}$ about the rotation axis (Fig. 5.2(a)). The unit vectors along orthogonal axes parallel to the north, east, and vertically downward directions at the position of the object are ($\mathbf{e}_N$, $\mathbf{e}_E$, $\mathbf{e}_D$) and define a local coordinate system. The horizontal velocity of the body has components (v_N, v_E, 0) parallel to these axes. The angular velocity of rotation has a constant direction. Transposed to the position of the moving body, it acts normal to the easterly component and has a positive northerly component at all latitudes. However, because $\mathbf{e}_D$ is defined to be positive downward, the vertical component is negative (upward) in the northern hemisphere and positive (downward) in the southern hemisphere. Thus the components of the rotation vector in the northern hemisphere are (ω_N, 0, $-\omega_D$). The velocity and rotation vectors are

$$\mathbf{v} = v_N\mathbf{e}_N + v_E\mathbf{e}_E \tag{5.13}$$

$$\boldsymbol{\omega} = \omega_N \mathbf{e}_N - \omega_D \mathbf{e}_D \tag{5.14}$$

Equation (5.12) can be evaluated by writing the vector cross product as a determinant:

$$\mathbf{a}_C = 2(\mathbf{v} \times \boldsymbol{\omega}) = 2 \begin{vmatrix} \mathbf{e}_N & \mathbf{e}_E & \mathbf{e}_D \\ v_N & v_E & 0 \\ \omega_N & 0 & -\omega_D \end{vmatrix} \tag{5.15}$$

On evaluating the determinant, we get

$$\mathbf{a}_C = 2(-v_E \omega_D \mathbf{e}_N + v_N \omega_D \mathbf{e}_E - v_E \omega_N \mathbf{e}_D) \tag{5.16}$$

In a geographic frame, the Coriolis acceleration has a component parallel to the vertical axis $\mathbf{e}_D$ and a component in the horizontal plane defined by $\mathbf{e}_N$ and $\mathbf{e}_E$.

5.2.1 Vertical component: the Eötvös effect

The last term in (5.16) describes the vertical component of the Coriolis acceleration:

$$\mathbf{a}_{E\ddot{o}} = -2v_E \omega_N \mathbf{e}_D \tag{5.17}$$

The formation of a vertical acceleration through the interaction of a horizontal east–west velocity with Earth's rotation is known as the *Eötvös effect*. It modifies the value of gravity measured from a moving platform, such as a vehicle, ship, or aircraft. If the body has an eastward velocity component (i.e., v_E is positive), $\mathbf{a}_{E\ddot{o}}$ acts in the direction of $-\mathbf{e}_D$, i.e., upwards. Conversely, if the velocity has a westward component, the Eötvös acceleration is downward. Its magnitude is dependent on the velocity and on the latitude through the value of ω_N, which is maximum at the equator and zero at the poles. For example, in a ship moving westwards at 7 knots (13 km hr^{-1}) at latitude 30 °N, the Eötvös acceleration increases the measured gravity by about 45 mgal. This greatly exceeds the measurement sensitivity in a marine gravity survey and necessitates a so-called *Eötvös correction* to gravity measurements.

5.2.2 Horizontal component: the Coriolis effect

The first two terms in (5.16) describe the horizontal component of the Coriolis acceleration:

$$\mathbf{a}_H = 2\omega_D(-v_E \mathbf{e}_N + v_N \mathbf{e}_E) \tag{5.18}$$

Its direction is normal to the velocity of the moving body, as can be verified by taking the scalar product of $\mathbf{a}_H$ and $\mathbf{v}$, which is zero:

$$(\mathbf{a}_H \cdot \mathbf{v}) = 2\omega_D(-v_E\mathbf{e}_N + v_N\mathbf{e}_E) \cdot (v_N\mathbf{e}_N + v_E\mathbf{e}_E) = 0 \qquad (5.19)$$

The angular velocity of rotation has a constant direction. Its vertical component ω_D is negative (upward) in the northern hemisphere and positive (downward) in the southern hemisphere. As a result, the Coriolis acceleration acts to the right of the direction of motion in the northern hemisphere, as can be seen by inspection of Fig. 5.2(b); it acts to the left in the southern hemisphere. The Coriolis effect causes deflection of the motion of bodies, such as air masses, moving across the surface of the Earth. In meteorology it gives rise to cyclonic and anticyclonic wind systems.

5.3 Precession and forced nutation of Earth's rotation axis

The main components of the precession and nutation result from the gravitational torques of the Sun and Moon on the Earth. In addition, the Sun's attraction causes the Moon's orbit to precess around the equator with a period of 18.6 yr. This motion results in a contribution to the nutation of Earth's rotation axis, which will be considered later. We first evaluate the precession and nutation caused by the solar torque, then extend the analysis to the lunar torque.

5.3.1 Effects of the torque due to the Sun's attraction

As the Earth moves around its orbit it experiences a variable torque due to the gravitational attraction of the Sun (Fig. 5.3(a)). For convenience assume that the Sun is at the center of the elliptical orbit. The tilt of the rotation axis inclines the northern hemisphere towards the Sun at the summer solstice and away from it at the winter solstice. Consider the Sun's attraction at the summer solstice (Fig. 5.3(b)). The gravitational attraction $\mathbf{F}_1$ on the part of the equatorial bulge closest to the Sun is greater than the attraction $\mathbf{F}_2$ on the opposite side. These forces are not collinear: the center of action of $\mathbf{F}_1$ is above the ecliptic, whereas that of $\mathbf{F}_2$ is below the ecliptic. The resulting torque $\mathbf{T}$ tries to reduce the tilt of the rotation axis. This causes the angular momentum vector to precess (Fig. 5.4(a)).

The torque causes an incremental change in angular momentum, $\Delta\mathbf{h}$, so that the angular momentum vector is displaced (Fig. 5.4(b)). Successive positions of the angular momentum vector lie on the surface of a cone whose axis is the pole to the ecliptic. The gravitational torque acts about an axis parallel to the line of

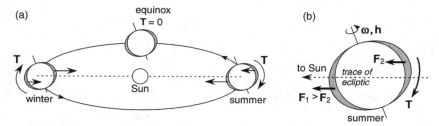

Fig. 5.3. (a) A torque of variable magnitude but constant direction is exerted by the Sun on the spinning Earth as it moves around its orbit. (b) A section through the inclined Earth in a plane normal to the ecliptic that includes the direction to the Sun, showing how the solar torque arises from unequal gravitational attraction on the equatorial bulge.

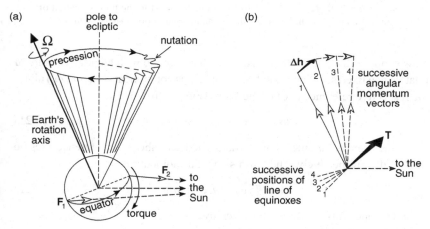

Fig. 5.4. (a) Precessional motion of the rotation axis about the pole to the ecliptic, on which nutation of the axis is superposed. (b) Incremental displacements of the angular momentum vector define the surface of a cone whose axis is the pole to the ecliptic. After Lowrie (2007).

equinoxes, which in turn is perpendicular to the rotation axis. As the angular momentum vector creeps over the surface of the cone, the line of equinoxes perpendicular to it moves around the ecliptic plane. The sense of motion is retrograde, opposite to the direction of the Earth's rotation.

Let x-, y-, and z-axes be defined as the orthogonal reference axes of the Earth's figure, with the z-axis parallel to the Earth's spin and the x–y plane coincident with the equatorial plane (Fig. 5.5(a)). The spin vector is

$$\mathbf{s} = s\mathbf{e}_z \qquad (5.20)$$

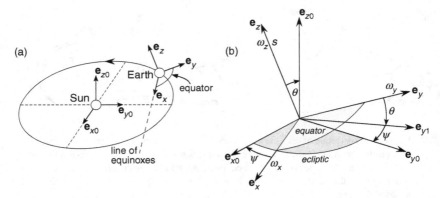

Fig. 5.5. (a) Definition of orthogonal reference axes relative to the Earth ($\mathbf{e}_x$, $\mathbf{e}_y$, $\mathbf{e}_z$) and to the ecliptic ($\mathbf{e}_{x0}$, $\mathbf{e}_{y0}$, $\mathbf{e}_{z0}$). (b) Rotations involved in the transformation of vector components from Earth coordinates to the Sun's coordinate system.

Now suppose that the reference axes are able to rotate with angular velocity $\boldsymbol{\omega}$ relative to a fixed set of coordinates, so that it has components (ω_x, ω_y, ω_z) along the respective reference axes of the Earth (Fig. 5.5(b)). Thus

$$\boldsymbol{\omega} = \omega_x \mathbf{e}_x + \omega_y \mathbf{e}_y + \omega_z \mathbf{e}_z \qquad (5.21)$$

Let the principal moments of inertia of the Earth about the reference axes be A, B, and C, respectively. The Earth's angular momentum is

$$\mathbf{h} = h_x \mathbf{e}_x + h_y \mathbf{e}_y + h_z \mathbf{e}_z \qquad (5.22)$$

The components (h_x, h_y, h_z) are given by

$$\begin{aligned} h_x &= A\omega_x \\ h_y &= B\omega_y \\ h_z &= C(s + \omega_z) \end{aligned} \qquad (5.23)$$

where h_z includes both the Earth's own spin and the z-component of the rotating coordinate system. The angular momentum is

$$\mathbf{h} = A\omega_x \mathbf{e}_x + B\omega_y \mathbf{e}_y + C(\omega_z + s)\mathbf{e}_z \qquad (5.24)$$

A torque $\mathbf{T}$ with components (L, M, N) along the respective reference axes causes a change of angular momentum given by

$$\mathbf{T} = \frac{d}{dt}\mathbf{h} = \frac{\partial}{\partial t}\mathbf{h} + (\boldsymbol{\omega} \times \mathbf{h}) \qquad (5.25)$$

The operator defined in (5.6) is used here to take into account the effect of Earth's rotation.

Using the determinant of components

$$\boldsymbol{\omega} \times \mathbf{h} = \begin{vmatrix} \mathbf{e}_x & \mathbf{e}_y & \mathbf{e}_z \\ \omega_x & \omega_y & \omega_z \\ h_x & h_y & h_z \end{vmatrix} \tag{5.26}$$

we obtain for the cross product

$$\boldsymbol{\omega} \times \mathbf{h} = \left(\omega_y h_z - \omega_z h_y \right) \mathbf{e}_x + \left(\omega_z h_x - \omega_x h_z \right) \mathbf{e}_y + \left(\omega_x h_y - \omega_y h_x \right) \mathbf{e}_z \tag{5.27}$$

Each of the x-, y-, and z-components of the motion described by (5.23) may now be analyzed in turn. For example, the x-component is

$$L = \frac{\partial h_x}{\partial t} + \left(\omega_y h_z - \omega_z h_y \right) \tag{5.28}$$

For succinctness we use the short form $\dot{\omega}_x = \partial \omega_x / \partial t$ in the following time-differentiations. We assume that the principal moments of inertia (A, B, C) are constant and that the changes in angular momentum result only from changes in angular rotation. Using the expressions in (5.23) for the components of angular momentum (h_x, h_y, h_z), we get

$$L = A\dot{\omega}_x + C\omega_y(\omega_z + s) - B\omega_y\omega_z \tag{5.29}$$

The equations of motion for the y- and z-components of the torque, M and N, are obtained in similar fashion and give the following:

$$M = B\dot{\omega}_y - C\omega_x(\omega_z + s) + A\omega_z\omega_x \tag{5.30}$$

$$N = C(\dot{\omega}_z + \dot{s}) + (B - A)\omega_x\omega_y \tag{5.31}$$

For the spheroidal Earth, the moments of inertia about all axes in the equatorial plane are equal, thus $A = B$ and (5.31) becomes

$$N = C(\dot{\omega}_z + \dot{s}) \tag{5.32}$$

As explained above, the gravitational torque of the Sun acts parallel to the line of equinoxes, and thus normal to the rotation axis. It has no component along the rotation axis, i.e., $N = 0$. Thus,

$$\dot{\omega}_z + \dot{s} = 0 \tag{5.33}$$

and

$$\omega_z + s = \Omega \tag{5.34}$$

where Ω is a constant rate of rotation. The remaining equations of motion can now be written as

$$L = A\dot{\omega}_x + C\omega_y\Omega - A\omega_y\omega_z \tag{5.35}$$

$$M = A\dot{\omega}_y - C\omega_x\Omega + A\omega_z\omega_x \tag{5.36}$$

The torque components L and M result from the gravitational attraction of the Sun on the spheroidal Earth (Fig. 5.3) and vary with the orbital position of the Earth, which is defined relative to the fixed axes. The angular velocity components are defined relative to Earth's reference axes, which are free to rotate. To solve the equations of motion it is necessary to establish a relationship between the fixed and rotating coordinate systems. The Sun's torque on the Earth must be derived and its components L and M along the rotating axes resolved.

5.3.2 Comparison of vectors in the coordinate systems of Earth and Sun

Let $(\mathbf{e}_{x0}, \mathbf{e}_{y0}, \mathbf{e}_{z0})$ be the orthogonal unit vectors of a solar coordinate system, defined so that $\mathbf{e}_{z0}$ is the pole to the ecliptic, $\mathbf{e}_{x0}$ is parallel to the minor axis, and $\mathbf{e}_{y0}$ is parallel to the major axis of Earth's elliptical orbit. Let $(\mathbf{e}_x, \mathbf{e}_y, \mathbf{e}_z)$ be orthogonal unit vectors for the rotating Earth, such that $\mathbf{e}_z$ is parallel to the spin axis and $\mathbf{e}_x$ lies along the intersection of the equatorial plane with the ecliptic, i.e., the line of equinoxes (Fig. 5.5(a)). The angle θ between $\mathbf{e}_z$ and $\mathbf{e}_{z0}$ is the obliquity of the rotation axis, and the angle ψ between $\mathbf{e}_x$ and $\mathbf{e}_{x0}$ defines the position of the line of equinoxes in the ecliptic plane.

The transformation of vector components from the Earth's coordinates to the Sun's coordinate system can be achieved with two rotations (Fig. 5.5(b)). The first is a rotation of θ about the x-axis. This aligns the rotation axis with the pole to the ecliptic, and brings $\mathbf{e}_y$ into an intermediate orientation $\mathbf{e}_{y1}$ in the ecliptic. The x-components of a vector are unchanged by this rotation. On comparing vector components we see that

$$\begin{aligned}\mathbf{e}_{y1} &= \mathbf{e}_y \cos\theta - \mathbf{e}_z \sin\theta \\ \mathbf{e}_{z0} &= \mathbf{e}_y \sin\theta + \mathbf{e}_z \cos\theta\end{aligned} \tag{5.37}$$

A second rotation of ψ about the pole to the ecliptic aligns $\mathbf{e}_x$ with $\mathbf{e}_{x0}$ and $\mathbf{e}_y$ with $\mathbf{e}_{y0}$. The $\mathbf{e}_{z0}$-components are not changed by this rotation, which gives the equations

$$\begin{aligned}\mathbf{e}_{x0} &= \mathbf{e}_x \cos\psi - \mathbf{e}_{y1} \sin\psi \\ \mathbf{e}_{y0} &= \mathbf{e}_x \sin\psi + \mathbf{e}_{y1} \cos\psi\end{aligned} \tag{5.38}$$

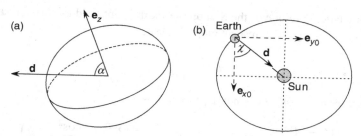

Fig. 5.6. (a) Definition of the angle α between the Earth's rotation axis $\mathbf{e}_z$ and the radial direction $\mathbf{d}$ to the Sun. (b) Definition of the angular orbital position χ of the Earth and the reference axes $\mathbf{e}_{x0}$ and $\mathbf{e}_{y0}$ in the ecliptic plane.

Substituting from (5.37) into (5.38) gives

$$\mathbf{e}_{x0} = \mathbf{e}_x \cos \psi - (\mathbf{e}_y \cos \theta - \mathbf{e}_z \sin \theta)\sin \psi \qquad (5.39)$$

$$\mathbf{e}_{y0} = \mathbf{e}_x \sin \psi + (\mathbf{e}_y \cos \theta - \mathbf{e}_z \sin \theta)\cos \psi \qquad (5.40)$$

After arranging terms, we get a set of equations relating the unit vectors ($\mathbf{e}_{x0}$, $\mathbf{e}_{y0}$, $\mathbf{e}_{z0}$) in the fixed coordinate system to the unit vectors ($\mathbf{e}_x$, $\mathbf{e}_y$, $\mathbf{e}_z$) in the rotating coordinate system:

$$\begin{aligned} \mathbf{e}_{x0} &= \mathbf{e}_x \cos \psi - \mathbf{e}_y \cos \theta \sin \psi + \mathbf{e}_z \sin \theta \sin \psi \\ \mathbf{e}_{y0} &= \mathbf{e}_x \sin \psi + \mathbf{e}_y \cos \theta \cos \psi - \mathbf{e}_z \sin \theta \cos \psi \\ \mathbf{e}_{z0} &= \mathbf{e}_y \sin \theta + \mathbf{e}_z \cos \theta \end{aligned} \qquad (5.41)$$

5.3.3 Computation of the Sun's torque on the Earth

The Sun's torque can be computed from the potential energy of the Earth–Sun pair. Let the angle between the Earth's rotation axis and the radial direction to the Sun at distance d be α (Fig. 5.6(a)). The gravitational potential U_G of the Earth at the Sun's location is obtained from the MacCullagh formula (Section 2.5),

$$U_G = -G\frac{M}{d} + G\frac{C - A}{d^3} P_2(\cos \alpha) \qquad (5.42)$$

Multiplying by the mass S of the Sun gives the potential energy U_{PE} of the gravitational interaction of Sun and Earth:

$$U_{PE} = -G\frac{ES}{d} + G\frac{(C - A)S}{d^3} P_2(\cos \alpha) \qquad (5.43)$$

The gravitational torque of the Sun on the Earth is obtained by differentiating the potential energy with respect to the angle α,

$$T = -\frac{\partial}{\partial \alpha} U_{PE} \tag{5.44}$$

The first term in (5.43) does not depend on α, so

$$T = -G\frac{(C-A)S}{d^3}\frac{\partial}{\partial \alpha} P_2(\cos \alpha) = -G\frac{(C-A)S}{d^3}\frac{\partial}{\partial \alpha}\left(\frac{3\cos^2\alpha - 1}{2}\right) \tag{5.45}$$

$$T = 3G\frac{(C-A)S}{d^3}\cos \alpha \sin \alpha \tag{5.46}$$

The Sun's torque on the equatorial bulge depends on the difference between the principal moments of inertia $(C-A)$, which would not exist for a spherical Earth. The torque depends on the angle α between the rotation axis $\mathbf{e}_z$ and the radius vector $\mathbf{d}$ from the Earth to the Sun, which varies as the Earth moves around its orbit. From Fig. 5.6(a) the following relationships are obtained:

$$(\mathbf{d} \cdot \mathbf{e}_z) = d\cos \alpha \tag{5.47}$$

$$(\mathbf{d} \times \mathbf{e}_z) = d\sin \alpha \tag{5.48}$$

The cross product $(\mathbf{d} \times \mathbf{e}_z)$ gives the correct sense of the torque of the Sun on the Earth. We can now substitute for $\sin \alpha$ and $\cos \alpha$ in (5.46), obtaining

$$\mathbf{T} = 3G\frac{(C-A)S}{d^5}(\mathbf{d} \cdot \mathbf{e}_z)(\mathbf{d} \times \mathbf{e}_z) \tag{5.49}$$

5.3.4 Equations of solar-induced precession and nutation

Referring to Fig. 5.6(b), the radial vector $\mathbf{d}$ can be written

$$\mathbf{d} = (d\cos\chi)\mathbf{e}_{x0} + (d\sin\chi)\mathbf{e}_{y0} \tag{5.50}$$

If the Earth orbits the Sun with constant angular velocity p, then in time t the radius vector moves through an angle $\chi = pt$. Therefore

$$\mathbf{d} = d(\mathbf{e}_{x0}\cos(pt) + \mathbf{e}_{y0}\sin(pt)) \tag{5.51}$$

The scalar product of $\mathbf{d}$ and $\mathbf{e}_z$ is

$$(\mathbf{d} \cdot \mathbf{e}_z) = d\cos(pt)(\mathbf{e}_{x0} \cdot \mathbf{e}_z) + d\sin(pt)(\mathbf{e}_{y0} \cdot \mathbf{e}_z) \tag{5.52}$$

We now substitute the expressions for e_{x0} and e_{y0} from (5.39) and (5.40), respectively, keeping in mind the following orthogonal relations between the unit vectors:

$$(e_x \cdot e_z) = (e_y \cdot e_z) = 0; \qquad (e_z \cdot e_z) = 1 \qquad (5.53)$$

This gives

$$(e_{x0} \cdot e_z) = (e_x \cos \psi - e_y \cos \theta \sin \psi + e_z \sin \theta \sin \psi) \cdot e_z$$
$$= \sin \theta \sin \psi \qquad (5.54)$$

$$(e_{y0} \cdot e_z) = (e_x \sin \psi + e_y \cos \theta \cos \psi - e_z \sin \theta \cos \psi) \cdot e_z$$
$$= -\sin \theta \cos \psi \qquad (5.55)$$

Inserting (5.54) and (5.55) into (5.52) gives

$$(\mathbf{d} \cdot e_z) = d \cos(pt) \sin \theta \sin \psi - d \sin(pt) \sin \theta \cos \psi$$
$$= -d \sin \theta \sin(pt - \psi) \qquad (5.56)$$

In order to determine the cross product

$$(\mathbf{d} \times e_z) = d \cos(pt)(e_{x0} \times e_z) + d \sin(pt)(e_{y0} \times e_z) \qquad (5.57)$$

we again make use of the orthogonality of the unit vectors:

$$(e_x \times e_z) = -e_y; \qquad (e_y \times e_z) = e_x; \qquad (e_z \times e_z) = 0 \qquad (5.58)$$

By again substituting for e_{x0} and e_{y0} from (5.39) and (5.40) we get

$$(e_{x0} \times e_z) = (e_x \cos \psi - e_y \cos \theta \sin \psi + e_z \sin \theta \sin \psi) \times e_z$$
$$= (e_x \times e_z) \cos \psi - (e_y \times e_z) \cos \theta \sin \psi$$
$$= -e_y \cos \psi - e_x \cos \theta \sin \psi \qquad (5.59)$$

$$(e_{y0} \times e_z) = (e_x \sin \psi + e_y \cos \theta \cos \psi - e_z \sin \theta \cos \psi) \times e_z$$
$$= (e_x \times e_z) \sin \psi + (e_y \times e_z) \cos \theta \cos \psi$$
$$= -e_y \sin \psi + e_x \cos \theta \cos \psi \qquad (5.60)$$

and, on inserting these expressions into (5.57), we have

$$(\mathbf{d} \times e_z) = - d \cos(pt)(e_y \cos \psi + e_x \cos \theta \sin \psi)$$
$$+ d \sin(pt)(-e_y \sin \psi + e_x \cos \theta \cos \psi) \qquad (5.61)$$

This equation can be simplified further by making use of trigonometric identities for the sine and cosine of the difference of two angles:

$$(\mathbf{d} \times \mathbf{e}_z) = d\cos\theta(\sin(pt)\cos\psi - \cos(pt)\sin\psi)\mathbf{e}_x$$
$$- d(\cos(pt)\cos\psi + \sin(pt)\sin\psi)\mathbf{e}_y \tag{5.62}$$

$$(\mathbf{d} \times \mathbf{e}_z) = d(\cos\theta\sin(pt - \psi)\mathbf{e}_x + \cos(pt - \psi)\mathbf{e}_y) \tag{5.63}$$

By combining the results for the scalar product (5.56) and cross product (5.63) we get the final expressions for the torque components L and M along the x- and y-axes, respectively:

$$L = -3G\frac{(C - A)S}{d^5}d^2 \sin\theta\cos\theta\sin^2(pt - \psi)$$
$$= -3G\frac{(C - A)S}{2d^3}\sin\theta\cos\theta(1 - \cos(2(pt - \psi))) \tag{5.64}$$

$$M = 3G\frac{(C - A)S}{d^5}d^2 \sin\theta\sin(pt - \psi)\cos(pt - \psi)$$
$$= 3G\frac{(C - A)S}{2d^3}\sin\theta\sin(2(pt - \psi)) \tag{5.65}$$

Upon inserting the equations for L and M into (5.35) and (5.36) we get

$$A\dot{\omega}_x + C\omega_y\Omega - A\omega_y\omega_z = -3G\frac{(C - A)S}{2d^3}\sin\theta\cos\theta(1 - \cos(2(pt - \psi)))$$
$$\tag{5.66}$$

$$A\dot{\omega}_y - C\omega_x\Omega + A\omega_x\omega_z = 3G\frac{(C - A)S}{2d^3}\sin\theta\sin(2(pt - \psi)) \tag{5.67}$$

5.3.5 Simplification of the equations of motion

The equations describe a forced harmonic motion, with the driving force dependent on the sine and cosine of $2(pt - \psi)$. It is easier to proceed with the solution of the equations if we simplify them by comparing the magnitudes of the terms on the left-hand side of each equation. This allows us to neglect terms that are unimportant to first order. Let the sine and cosine functions be represented by the real and imaginary parts of a complex number (Section 1.2) with phase equal to $2(pt - \psi)$; we can write it as $\exp[2i(pt - \psi)]$. Each equation then has the form

$$a\dot{\omega} + b\omega + c\omega^2 \sim \exp[2i(pt - \psi)] \tag{5.68}$$

in which ω stands for either of the angular velocities ω_x and ω_y. The driving force on the right-hand side of the equation is periodic with angular frequency $2p$.

The solution of the equation must also be periodic, so we may expect that $|\dot{\omega}_x| \approx 2p\omega_x$ and $|\dot{\omega}_y| \approx 2p\omega_y$.

The rotation Ω of the Earth about its axis has period $2\pi/\Omega = 1$ day; the angular velocity p of the Earth about the Sun has period 365 days, so $\Omega = 365p$. The angular velocity components of the rotating coordinate system are much smaller than the daily rotation rate of the Earth: $\omega_x \sim \omega_y \ll \Omega$. On comparing the first and second terms on the left of (5.67) and (5.68) we see that the first term can be neglected because

$$|\dot{\omega}| \sim 2p\omega \ll \Omega\omega \tag{5.69}$$

Similarly, the magnitude of the third term may be neglected compared with the second term because

$$|\omega_y\omega_z| \sim \omega^2 \ll \Omega\omega \tag{5.70}$$

Thus $C\omega_x\Omega$ and $C\omega_y\Omega$ are the dominant terms on the left of the equations and the other terms on the left may be neglected by comparison. This leads to simpler equations of motion, such as

$$C\omega_x\Omega = -3G\frac{(C-A)S}{2d^3}\sin\theta\sin(2(pt-\psi)) \tag{5.71}$$

from which

$$\omega_x = -\frac{3GS}{2\Omega d^3}\left(\frac{C-A}{C}\right)\sin\theta\sin(2(pt-\psi)) \tag{5.72}$$

Similarly,

$$\omega_y = -\frac{3GS}{2\Omega d^3}\left(\frac{C-A}{C}\right)\sin\theta\cos\theta(1-\cos(2(pt-\psi))) \tag{5.73}$$

The angular velocities of the rotating coordinate axes are related to the rates of change with time of the angles θ and ψ. It is evident by reference to Fig. 5.5(b) that

$$\omega_x = \frac{\partial\theta}{\partial t}; \quad \omega_y = \sin\theta\frac{\partial\psi}{\partial t}; \quad \omega_z = \cos\theta\frac{\partial\psi}{\partial t} \tag{5.74}$$

The same parameters appear on the right of each equation of motion. We can substitute

$$F_S = -\frac{3GS}{2\Omega d^3}\left(\frac{C-A}{C}\right) \tag{5.75}$$

Using these relationships, the equations of motion become

$$\frac{\partial \theta}{\partial t} = F_S \sin \theta \sin(2(pt - \psi)) \tag{5.76}$$

$$\frac{\partial \psi}{\partial t} = F_S \cos \theta - F_S \cos \theta \cos(2(pt - \psi)) \tag{5.77}$$

5.3.6 Precession and nutation induced by the Sun

The angle ψ defines the position of the line of equinoxes in the ecliptic plane. Equation (5.77) shows that the rate of change of ψ consists of two parts. The first term, $F_S \cos \theta$, describes a motion of the x-axis – the line of equinoxes – around the ecliptic plane, at a constant rate. The rotation axis (z-axis) moves accordingly, staying orthogonal to the x-axis. The rotation axis thus moves across the surface of a cone whose axis is the pole to the ecliptic (Fig. 5.4(a)). This motion is the precession of the rotation axis. The mean precession rate is 50.385 arcsec per year, corresponding to a period of 25,720 yr. The term F_S is negative (5.75), so the precession is retrograde, i.e., the motion is in the opposite sense to Earth's rotation. The parameters that define F_S have constant values, all of which are known except the moments of inertia, A and C. The ratio H defined by

$$H = \frac{C - A}{C} \tag{5.78}$$

is the dynamic ellipticity of the Earth. It can be calculated from the observed rate of precession and has the value $3.273\,787\,5 \times 10^{-3}$ (1/305.457).

The term on the right of (5.76) describes a periodic fluctuation in the obliquity θ. This "nodding" motion is called the *nutation in obliquity* of the rotation axis. A similar fluctuation of the angle ψ is shown by the second term on the right of (5.77). This fluctuation occurs in the plane of the ecliptic and is known as the *nutation in longitude*. These forced nutations each have the same frequency, $2p$, corresponding to a period of half a year (183 days). They are called the semi-annual nutations. Their amplitudes are very small and unequal, amounting to only a few seconds of arc. Using for convenience the short form for time-differentiations, we can write

$$\frac{\dot{\theta}}{F_S \sin \theta} = \sin(2(pt - \psi)) \tag{5.79}$$

$$\frac{\dot{\psi} - F_S \cos \theta}{F_S \cos \theta} = -\cos(2(pt - \psi)) \tag{5.80}$$

Squaring both sides and summing gives

$$\frac{(\dot{\psi} - F_S \cos \theta)^2}{(F_S \cos \theta)^2} + \frac{(\dot{\theta})^2}{(F_S \sin \theta)^2} = 1 \tag{5.81}$$

The equation of an ellipse with semi-major axes a and b is

$$\frac{x^2}{a^2} + \frac{y^2}{b^2} = 1 \tag{5.82}$$

On comparing (5.79) and (5.80) we see that the two forced nutations combine to produce an elliptical motion of the rotation axis about its mean position, superposed on the steady motion around the precession cone (Fig. 5.4(a)).

5.3.7 Precession and nutation induced by the Moon

The Earth's nearest neighbor, the Moon, is much smaller than the distant Sun, but its gravitational effect also causes both precession and nutation of the Earth's rotation axis. The combined effects of Sun and Moon are known as the *lunisolar* precession and nutation. The effects of the attraction of the Moon's mass M on Earth's equatorial bulge are analyzed in the same way as the solar torque, and we get equations that have the same form as (5.76) and (5.77). Using subscript L to identify the lunar parameters, we get

$$\dot{\theta}_L = F_L \sin \theta_L \sin(2(p_L t - \psi_L)) \tag{5.83}$$

$$\dot{\psi}_L = F_L \cos \theta_L - F_L \cos \theta_L \cos(2(p_L t - \psi_L)) \tag{5.84}$$

Here the angles θ_L and ψ_L locate the rotation axis relative to the Moon's orbit, and p_L is the angular velocity of the Moon around the Earth. This gives a nutation component with a period of half a month. Because the Moon's orbit is only slightly inclined to the ecliptic, the solar and lunar effects can be added as scalars.

The constant F_L depends on the mass M of the Moon and its distance d_L from the Earth:

$$F_L = -\frac{3GM}{2\Omega d_L^3} \left(\frac{C - A}{C}\right) \tag{5.85}$$

It is interesting to compare this term for the lunar effect with the corresponding term for the Sun's influence on the precession (using subscript S for the respective solar parameters):

$$\frac{F_L}{F_S} = \frac{-\dfrac{3GM}{2\Omega d_L^3}\left(\dfrac{C-A}{C}\right)}{-\dfrac{3GS}{2\Omega d_S^3}\left(\dfrac{C-A}{C}\right)} = \left(\frac{M}{S}\right)\left(\frac{d_S}{d_L}\right)^3 \tag{5.86}$$

The masses of the Sun and Moon and their distances from the Earth are given in Table 4.1. Inserting the appropriate values gives

$$\frac{F_L}{F_S} = \left(\frac{M}{S}\right)\left(\frac{d_S}{d_L}\right)^3 = 2.2 \tag{5.87}$$

The ratio is the same as that involved in comparing the tide-raising accelerations of the Sun and Moon (Section 4.2.3), and the explanation of the result is the same. The mass of the Moon is much smaller than that of the Sun, but the ratio of their influences depends on the cube of the distance ratio, so the Moon accounts for about two thirds of the combined lunisolar precession and nutation, and the Sun about one third.

5.3.8 Nutation due to precession of the Moon's orbit

As a result of tidal friction the Moon's spin rate about its own axis is the same as its orbital angular velocity p_L about the Earth. If the moment of inertia of the Moon about its spin axis is I_L, its mass M and radius R_L (1,738 km), the spin angular momentum is

$$h_L = I_L p_L = k_L M R_L^2 p_L \tag{5.88}$$

For the Moon k_L is equal to 0.394. For a uniform sphere $k_L = 0.4$. A smaller value indicates that density increases with depth, e.g., for the Earth $k_E = 0.3308$.

The orbital angular momentum is

$$h_O = M r_L^2 p_L \tag{5.89}$$

where r_L is the radius of the Moon's orbit (384,400 km)

On comparing the spin and orbital angular momenta, we have

$$\frac{h_L}{h_O} = \frac{k_L M R_L^2 p_L}{M r_L^2 p_L} = k_L \left(\frac{R_L}{r_L}\right)^2 \tag{5.90}$$

Upon inserting appropriate values, it is evident that the Moon's spin angular momentum is much less than its orbital angular momentum.

The Moon's orbit and its angular momentum vector are inclined at a small angle (5.145°) to the ecliptic plane. The Sun's attraction results in a torque that attempts to turn the inclined angular momentum vector normal to the ecliptic.

Similarly to the effect of the Sun on Earth's angular momentum (Fig. 5.4(b)), the solar torque causes the Moon's orbit to precess about the pole to the ecliptic. The effective inclination of the Moon's orbit to the Earth's rotation axis varies between 18.28° and 28.58° (i.e., 23.43 ± 5.15°) with a period of 18.6 yr, which results in a corresponding component in the nutation of Earth's rotation axis. The precession of the Moon's orbit causes the largest part of the nutation, with amplitudes of 9.2 arcsec in obliquity and 17.3 arcsec in longitude. The semi-annual nutation has amplitudes of only 1.3 arcsec in longitude and 0.6 arcsec in obliquity.

5.4 The free, Eulerian nutation of a rigid Earth

External forces on the spinning Earth give rise to the forced nutation and precession of the rotation axis. These were described by allowing the reference axes of the Earth to rotate relative to the spin axis. The long-term average rotation of the Earth gives it a spheroidal shape about the axis of figure. If a symmetric body spins freely about its axis of symmetry, its orientation in space remains fixed. However, if some event displaces the spin axis from its mean direction, the Earth's instantaneous rotation is no longer about its axis of symmetry. This results in a motion called the *free nutation*. It was predicted in the eighteenth century by the Swiss mathematician Leonhard Euler, and is also called *Eulerian nutation*. The use of the term nutation is an unfortunate misnomer as the motion does not involve "nodding" of the spin axis. In Eulerian nutation the instantaneous rotation axis moves around the surface of a cone whose axis is the axis of symmetry.

Let the reference axes be defined relative to the figure of the Earth so that the z-axis agrees with the axis of symmetry and the x- and y-axes lie in the equatorial plane (Fig. 5.7). The reference axes rotate along with the Earth, so the angular

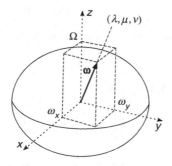

Fig. 5.7. Angular velocity components $(\omega_x, \omega_y, \Omega)$ and direction cosines (λ, μ, ν) of the displaced instantaneous rotation axis.

velocity ω_z about the z-axis is the same as the Earth's spin Ω. A displacement of the instantaneous spin vector is represented by angular velocities ω_x and ω_y about the equatorial axes. The instantaneous rotation vector is then

$$\boldsymbol{\omega} = \omega_x \mathbf{e}_x + \omega_y \mathbf{e}_y + \omega_z \mathbf{e}_z \tag{5.91}$$

Using as before A, B, and C for the principal moments of inertia about the x-, y-, and z-axes, respectively, the angular momentum is given by

$$\mathbf{h} = A\omega_x \mathbf{e}_x + B\omega_y \mathbf{e}_y + C\omega_z \mathbf{e}_z \tag{5.92}$$

In contrast to the forced motion of the rotation axis caused by solar and lunar attraction, the motion of the rotation axis is in this case free of external torques. Thus

$$\mathbf{T} = \frac{d\mathbf{h}}{dt} = \frac{\partial \mathbf{h}}{\partial t} + (\boldsymbol{\omega} \times \mathbf{h}) = 0 \tag{5.93}$$

Assuming that the Earth rotates as a rigid body, the equations of motion for each of the reference axes can be developed as in the case of forced nutation (see Section 5.3.1):

$$\begin{aligned} A\dot{\omega}_x + (C - B)\omega_y\omega_z &= 0 \\ B\dot{\omega}_y + (A - C)\omega_x\omega_z &= 0 \\ C\dot{\omega}_z + (B - A)\omega_x\omega_y &= 0 \end{aligned} \tag{5.94}$$

The symmetry of the Earth's figure implies that the equatorial moments of inertia are equal, $A = B$:

$$A\dot{\omega}_x + (C - A)\omega_y\omega_z = 0 \tag{5.95}$$

$$A\dot{\omega}_y - (C - A)\omega_x\omega_z = 0 \tag{5.96}$$

$$C\dot{\omega}_z = 0 \tag{5.97}$$

The last equation requires that the angular velocity about the z-axis is constant:

$$\omega_z = \Omega \tag{5.98}$$

Rewriting (5.95) and (5.96) gives

$$\dot{\omega}_x + \left(\frac{C - A}{A}\right)\Omega\omega_y = 0 \tag{5.99}$$

$$\dot{\omega}_y - \left(\frac{C - A}{A}\right)\Omega\omega_x = 0 \tag{5.100}$$

Differentiating (5.99) with respect to time t gives

$$\ddot{\omega}_x + \left(\frac{C-A}{A}\right)\Omega\dot{\omega}_y = 0 \tag{5.101}$$

We can now substitute from (5.100) into (5.101), which gives an equation for ω_x:

$$\ddot{\omega}_x + \left(\frac{C-A}{A}\right)^2 \Omega^2 \omega_x = 0 \tag{5.102}$$

This equation represents a simple harmonic motion and has the solution

$$\omega_x = \omega_0 \cos\left(\frac{C-A}{A}\Omega t + \delta\right) \tag{5.103}$$

where ω_0 is the amplitude and δ the phase. By substituting this result into (5.100) and solving for ω_y we get

$$\omega_y = \omega_0 \sin\left(\frac{C-A}{A}\Omega t + \delta\right) \tag{5.104}$$

Equations (5.103) and (5.104) describe a periodic motion of the instantaneous spin axis about the axis of figure. It is called the free nutation (or Euler nutation). Its period is

$$\tau_0 = \frac{2\pi}{\Omega}\left(\frac{A}{C-A}\right) \tag{5.105}$$

The factor $2\pi/\Omega$ represents the daily rotation of the Earth, so the period of the free nutation is $A/(C-A)$ days. The dynamic ellipticity obtained from the precession period (5.78) indicates that this period is about 305 days (~10 months). However, astronomers in the eighteenth and early nineteenth centuries were unable to detect a motion of Earth's axis with this period. The reason lies in the assumption that the Earth rotates as a rigid body. In fact its elasticity allows it to deform slightly as a result of the displacement of the instantaneous rotation axis from the axis of figure, and this extends the period to 435 days (~14 months). The observed motion is called the Chandler wobble.

5.5 The Chandler wobble

The Chandler wobble is a somewhat irregular cyclical motion of the instantaneous rotation axis with a period of about 435 days and an amplitude of a few

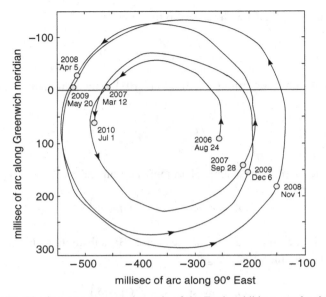

Fig. 5.8. The instantaneous rotation axis of the Earth exhibits a nearly circular motion with period 435 days – the Chandler wobble – and an annual circular motion. These motions are superposed on a slow drift of about 20 m per century along longitude 80 °W. Data source: International Earth Rotation and Reference Systems Service.

tenths of a second of arc, approximately 10–15 m (Fig. 5.8). The displacement of the rotation axis from its mean position is thought to result from changes in oceanic circulation and fluctuations in atmospheric pressure. The displacement of the instantaneous rotation axis from the axis of figure gives rise to an asymmetry in the Earth's shape. The moments of inertia A, B, and C about the reference axes are no longer adequate to describe the inertia tensor. The products of inertia H, J, and K are needed to express the asymmetry of the mass distribution (see Box 2.2). Let the instantaneous rotation axis have a direction specified by direction cosines (λ, μ, ν) relative to the x-, y-, and z-axes defined in Fig. 5.7. The moment of inertia I about the instantaneous rotation axis is given by (2.134):

$$I = A\lambda^2 + B\mu^2 + C\nu^2 - 2K\lambda\mu - 2H\mu\nu - 2J\nu\lambda$$

On writing $I_{11} = A$, $I_{22} = B$, and $I_{33} = C$ for the principal moments of inertia and $I_{12} = I_{21} = -K$, $I_{13} = I_{31} = -J$, and $I_{23} = I_{32} = -H$ for the products of inertia (Box 5.1), this equation becomes

$$I = I_{11}\lambda^2 + I_{22}\mu^2 + I_{33}\nu^2 + 2I_{12}\lambda\mu + 2I_{23}\mu\nu + 2I_{31}\nu\lambda \tag{5.106}$$

The angular velocity has components $(\omega_x, \omega_y, \Omega)$. Using numerical subscripts 1, 2, and 3 for the x-, y-, and z-components, respectively, the angular momentum h and angular velocity ω are related by the tensor equation

$$h_i = I_{ij}\omega_j \tag{5.107}$$

where the symmetric *inertia tensor* I_{ij} (Box 5.1) represents the elements of the matrix

---

Box 5.1. **The inertia tensor**

Let a rigid body be composed of elementary particles with mass m_i and coordinates (x_i, y_i, z_i) relative to an orthogonal Cartesian coordinate system. Let the body rotate with angular velocity ω about an axis through the origin. The linear velocity of a particle m_i at distance r_i from the origin is

$$\mathbf{v}_i = \omega \times \mathbf{r}_i \tag{1}$$

The linear momentum of the particle is $m_i v_i$ and its contribution to the angular momentum of the rotating body is

$$\mathbf{h}_i = \mathbf{r}_i \times m_i \mathbf{v}_i \tag{2}$$

The angular momentum of the body is

$$\mathbf{h} = \sum_i m_i(\mathbf{r}_i \times \mathbf{v}_i) = \sum_i m_i(\mathbf{r}_i \times (\omega \times \mathbf{r}_i)) \tag{3}$$

Using the identity in (1.18), the vector cross product is

$$\mathbf{r}_i \times (\omega \times \mathbf{r}_i) = \omega r_i^2 - \mathbf{r}_i(\omega \cdot \mathbf{r}_i) \tag{4}$$

On substituting this expression into (3), the angular momentum becomes

$$\mathbf{h} = \omega \sum_i m_i r_i^2 - \sum_i m_i \mathbf{r}_i(\omega \cdot \mathbf{r}_i) \tag{5}$$

The x-component h_x is

$$h_x = \omega_x \sum_i m_i(x_i^2 + y_i^2 + z_i^2) - \sum_i m_i x_i(\omega_x x_i + \omega_y y_i + \omega_z z_i) \tag{6}$$

$$h_x = \omega_x \sum_i m_i \left(y_i^2 + z_i^2\right) - \omega_y \sum_i m_i x_i y_i - \omega_z \sum_i m_i z_i x_i \qquad (7)$$

Analogously, the y- and z-components, h_y and h_z, of the angular momentum are, respectively,

$$h_y = -\omega_x \sum_i m_i y_i x_i + \omega_y \sum_i m_i \left(z_i^2 + x_i^2\right) - \omega_z \sum_i m_i y_i z_i \qquad (8)$$

$$h_z = -\omega_x \sum_i m_i z_i x_i - \omega_y \sum_i m_i z_i y_i + \omega_z \sum_i m_i \left(x_i^2 + y_i^2\right) \qquad (9)$$

Using the definitions of moments and products of inertia in Box 2.2, the angular momentum components are

$$\begin{aligned} h_x &= A\omega_x - K\omega_y - J\omega_z \\ h_y &= -K\omega_x + B\omega_y - H\omega_z \\ h_z &= -J\omega_x - H\omega_y + C\omega_z \end{aligned} \qquad (10)$$

These equations relating the components of **h** and **ω** can be written as a single matrix equation,

$$\begin{pmatrix} h_x \\ h_y \\ h_z \end{pmatrix} = \begin{pmatrix} A & -K & -J \\ -K & B & -H \\ -J & -H & C \end{pmatrix} \begin{pmatrix} \omega_x \\ \omega_y \\ \omega_z \end{pmatrix} \qquad (11)$$

Using numerical subscripts 1, 2, and 3 for the x-, y-, and z-components, respectively, the moments of inertia (diagonal elements) are represented by $I_{11} = A$, $I_{22} = B$, and $I_{33} = C$. The products of inertia (non-diagonal elements) are $I_{12} = I_{21} = -K$, $I_{13} = I_{31} = -J$, and $I_{23} = I_{32} = -H$. The matrix equation is then

$$\begin{pmatrix} h_1 \\ h_2 \\ h_3 \end{pmatrix} = \begin{pmatrix} I_{11} & I_{12} & I_{13} \\ I_{21} & I_{22} & I_{23} \\ I_{31} & I_{32} & I_{33} \end{pmatrix} \begin{pmatrix} \omega_1 \\ \omega_2 \\ \omega_3 \end{pmatrix} \qquad (12)$$

In tensor notation this equation is written succinctly as

$$h_i = I_{ij}\omega_j \quad (i = 1, 2, 3; j = 1, 2, 3) \qquad (13)$$

The symmetric, second-order tensor I_{ij}, whose components are the moments and products of inertia, is called the *inertia tensor*.

$$I_{ij} = \begin{pmatrix} I_{11} & I_{12} & I_{13} \\ I_{21} & I_{22} & I_{23} \\ I_{31} & I_{32} & I_{33} \end{pmatrix} \tag{5.108}$$

Equation (5.93) for the free motion of the displaced instantaneous rotation axis becomes

$$\dot{h}_i + (\boldsymbol{\omega} \times \mathbf{h})_i = 0 \tag{5.109}$$

Upon inserting (5.107), we have for the first term

$$\dot{h}_i = \frac{\partial}{\partial t}(I_{ij}\omega_j) = \dot{I}_{ij}\omega_j + I_{ij}\dot{\omega}_j \tag{5.110}$$

The x-, y-, and z-components of the cross product have the form

$$(\boldsymbol{\omega} \times \mathbf{h})_1 = \omega_2 I_{3k}\omega_k - \omega_3 I_{2k}\omega_k \tag{5.111}$$

The components of the equation of motion become

$$\begin{aligned} \dot{h}_1 + \omega_2 I_{3k}\omega_k - \omega_3 I_{2k}\omega_k &= 0 \\ \dot{h}_2 + \omega_3 I_{1k}\omega_k - \omega_1 I_{3k}\omega_k &= 0 \\ \dot{h}_3 + \omega_1 I_{2k}\omega_k - \omega_2 I_{1k}\omega_k &= 0 \end{aligned} \tag{5.112}$$

By expanding these equations of motion separately, we obtain expressions for each individual component.

For the x-component,

$$I_{11}\dot{\omega}_1 + I_{12}\dot{\omega}_2 + I_{13}\dot{\omega}_3 + \dot{I}_{11}\omega_1 + \dot{I}_{12}\omega_2 + \dot{I}_{13}\omega_3$$
$$+ \omega_2 I_{31}\omega_1 + \omega_2 I_{32}\omega_2 + \omega_2 I_{33}\omega_3 - \omega_3 I_{21}\omega_1 - \omega_3 I_{22}\omega_2 - \omega_3 I_{23}\omega_3 = 0 \tag{5.113}$$

For the y-component,

$$I_{21}\dot{\omega}_1 + I_{22}\dot{\omega}_2 + I_{23}\dot{\omega}_3 + \dot{I}_{21}\omega_1 + \dot{I}_{22}\omega_2 + \dot{I}_{23}\omega_3$$
$$+ \omega_3 I_{11}\omega_1 + \omega_3 I_{12}\omega_2 + \omega_3 I_{13}\omega_3 - \omega_1 I_{31}\omega_1 - \omega_1 I_{32}\omega_2 - \omega_1 I_{33}\omega_3 = 0 \tag{5.114}$$

For the z-component,

$$I_{31}\dot{\omega}_1 + I_{32}\dot{\omega}_2 + I_{33}\dot{\omega}_3 + \dot{I}_{31}\omega_1 + \dot{I}_{32}\omega_2 + \dot{I}_{33}\omega_3$$
$$+ \omega_1 I_{21}\omega_1 + \omega_1 I_{22}\omega_2 + \omega_1 I_{23}\omega_3 - \omega_2 I_{11}\omega_1 - \omega_2 I_{12}\omega_2 - \omega_2 I_{13}\omega_3 = 0 \tag{5.115}$$

5.5.1 Simplification of the equations of motion

Each of the equations of motion contains many terms, some of which are effectively irrelevant because they are very small compared with other terms. In order to obtain an analytical solution it is necessary to introduce some approximations, as follows.

1. The angular velocities (ω_1, ω_2) are small compared with the daily rotation Ω. We will retain ω_1 and ω_2 to first order but neglect their products and higher orders, i.e.,

$$\omega_1^2 = \omega_2^2 = \omega_1\omega_2 = 0$$

2. The products of inertia (non-diagonal elements in the inertia tensor) are small, and we may neglect their products with the velocities (ω_1, ω_2), i.e.,

$$I_{13}\omega_1 = I_{13}\omega_2 = I_{12}\omega_1 = I_{12}\omega_2 = I_{23}\omega_1 = I_{23}\omega_2 = 0$$

3. We may also assume that the products of inertia change very slowly with time. In this case we may neglect further products with the velocities (ω_1, ω_2), i.e.,

$$\dot{I}_{13}\omega_1 = \dot{I}_{13}\omega_2 = \dot{I}_{12}\omega_1 = \dot{I}_{12}\omega_2 = \dot{I}_{23}\omega_1 = \dot{I}_{23}\omega_2 = 0$$

4. We may assume that the principal moments of inertia A, B, and C do not change with time, i.e., only the asymmetry in the mass distribution is responsible for the wobble of the rotation axis. That is,

$$\dot{I}_{ii} = 0$$

If we now apply these assumptions to the equations of motion, most of the terms drop out. For example, (5.115) reduces to

$$I_{33}\dot{\omega}_3 = 0 \tag{5.116}$$

This leads to the same result as for the Euler precession of the rigid Earth, namely that the angular velocity about the axis of figure is constant:

$$\omega_3 = \Omega \tag{5.117}$$

The remaining two equations of motion reduce to

$$I_{11}\dot{\omega}_1 + \dot{I}_{13}\omega_3 + \omega_2\omega_3(I_{33} - I_{22}) - \omega_3^2 I_{23} = 0 \tag{5.118}$$

$$I_{22}\dot{\omega}_2 + \dot{I}_{23}\omega_3 + \omega_3\omega_1(I_{11} - I_{33}) + \omega_3^2 I_{13} = 0 \tag{5.119}$$

These can now be rewritten with the more easily recognizable parameters for the moments and products of inertia:

$$A\dot{\omega}_1 + \omega_2\Omega(C - A) + \Omega^2 H - \dot{J}\Omega = 0 \qquad (5.120)$$

$$A\dot{\omega}_2 - \omega_1\Omega(C - A) - \Omega^2 J - \dot{H}\Omega = 0 \qquad (5.121)$$

The displacement of the instantaneous axis of rotation from the z-axis is very small, amounting to less than 0.25 arcsec. The direction cosines of the rotation axis may therefore be written as $(\lambda, \mu, 1)$ and the angular velocities as $(\omega_1 = \lambda\Omega, \omega_2 = \mu\Omega)$. Upon inserting these values into the equations of motion and dividing throughout by Ω, we get the simultaneous equations

$$A\dot{\lambda} + \mu\Omega(C - A) + \Omega H - \dot{J} = 0 \qquad (5.122)$$

$$A\dot{\mu} - \lambda\Omega(C - A) - \Omega J - \dot{H} = 0 \qquad (5.123)$$

Note that the product of inertia K, which describes asymmetry in the x–y plane, does not play a role in the wobble equations. Only asymmetries in the y–z and z–x planes that include the rotation axis determine the wobble motion. This will become evident when we compute the values of the products of inertia H and J, which we will obtain from a comparison with the MacCullagh equation for the gravitational potential of the non-spheroidal Earth.

5.5.2 Computation of the products of inertia

The Earth is deformed by the centrifugal force of its rotation, the main result being its spheroidal shape. If the axis of rotation is displaced from the axis of symmetry of a rigid Earth, the spheroid exhibits Euler nutation about the spin axis without additional deformation (Fig. 5.9(a)). However, the body of an elastic Earth can adjust its shape to the displaced spin axis by deforming further, as illustrated in Fig. 5.9(b). Parts of the ellipsoid are elevated above the original spheroid (regions "e"), while other parts are depressed below it (regions "d"). The shape conforming to the elastic deformation caused by the Chandler wobble is not symmetric with respect to the reference axes. This gives rise to the products of inertia H and J.

At a point in the Earth specified by co-latitude θ and radial distance r the distance from the rotation axis is $r \sin \theta$ and the potential Φ of the centrifugal acceleration is

$$\Phi = -\frac{1}{2}\Omega^2 r^2 \sin^2\theta = -\frac{1}{2}\Omega^2 r^2 + \frac{1}{2}\Omega^2 r^2 \cos^2\theta \qquad (5.124)$$

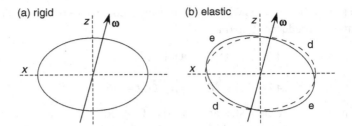

Fig. 5.9. (a) Displacement of the rotation axis of a *rigid* Earth results in Euler nutation without additional deformation. (b) The *elastic* Earth adjusts its shape to the displaced spin axis by deforming further, so that regions "e" lie above and regions "d" lie below the elliptical section (dashed) of the rigid body.

Let the Cartesian coordinates of the point be (x, y, z). The direction cosines $(\lambda_0, \mu_0, \nu_0)$ of the radius through the point at (r, θ) are

$$\lambda_0 = \frac{x}{r}; \quad \mu_0 = \frac{y}{r}; \quad \nu_0 = \frac{z}{r} \tag{5.125}$$

If the direction cosines of the instantaneous rotation axis are (λ, μ, ν), then θ is approximately the angle between the two lines, and

$$\cos \theta = \lambda\lambda_0 + \mu\mu_0 + \nu\nu_0 \tag{5.126}$$

Inserting the values from (5.125) gives

$$r \cos \theta = \lambda x + \mu y + \nu z \tag{5.127}$$

and, using this relationship in (5.124), we get the centrifugal potential

$$\Phi = -\frac{1}{2}\Omega^2 \left(x^2 + y^2 + z^2\right) + \frac{1}{2}\Omega^2 (\lambda x + \mu y + \nu z)^2 \tag{5.128}$$

$$\Phi = -\frac{1}{2}\Omega^2 \left(x^2 + y^2 + z^2\right)$$
$$+ \frac{1}{2}\Omega^2 \left(\lambda^2 x^2 + \mu^2 y^2 + \nu^2 z^2 + 2\lambda\mu xy + 2\mu\nu yz + 2\nu\lambda zx\right) \tag{5.129}$$

This may be simplified as before by setting the second-order values $\lambda^2 = \mu^2 = \lambda\mu = 0$ and $\nu^2 = \nu = 1$. Then

$$\Phi = -\frac{1}{2}\Omega^2 \left(x^2 + y^2\right) + \Omega^2 z(\lambda x + \mu y) \tag{5.130}$$

The first term here is the centrifugal potential due to rotation about the axis of figure. The second term is the extra centrifugal potential Φ_2 due to the displacement of the instantaneous rotation axis in the Chandler wobble,

$$\Phi_2 = \Omega^2 z(\lambda x + \mu y) \tag{5.131}$$

The wobble potential is a second-order solution of Laplace's equation, because

$$\nabla^2 \Phi_2 = \Omega^2 \left(\frac{\partial^2}{\partial x^2} + \frac{\partial^2}{\partial y^2} + \frac{\partial^2}{\partial z^2} \right)(\lambda zx + \mu yz) = 0 \tag{5.132}$$

Φ_2 is a deforming potential and causes a corresponding deformation that has its own gravitational potential Φ_i, which, as in the theory of the equilibrium tides, is proportional to Φ_2,

$$\Phi_i = k\Omega^2 z(\lambda x + \mu y) \tag{5.133}$$

The constant of proportionality k is the first Love number. The potential Φ_i is a solution of Laplace's equation for a space in which r can be zero. In our case it describes the wobble centrifugal potential within the Earth. We need a solution that is valid outside the Earth. As shown in Section 4.3.2 for the tidal gravity anomaly, the general solution Φ of Laplace's equation may be written

$$\Phi = \left(Ar^2 + \frac{B}{r^3} \right) P_2(\cos \theta) = \Phi_i + \Phi_e \tag{5.134}$$

where the first part Φ_i is valid inside and the second part Φ_e outside a volume of interest. The two solutions vary differently with radial distance r, but their ratio for the Earth with radius R is

$$\Phi_e = \left(\frac{R}{r} \right)^5 \Phi_i \tag{5.135}$$

On substituting for Φ_i from (5.133), the potential of the deformation caused by the wobble is

$$\Phi_e = \frac{R^5}{r^5} k\Omega^2 z(\lambda x + \mu y) \tag{5.136}$$

On converting the Cartesian coordinates (x, y, z) to direction cosines $(\lambda_0, \mu_0, \nu_0)$ of the line through the point of observation (5.125), we get the potential Φ_e of the wobble deformation at an external point:

$$\Phi_e = \frac{k\Omega^2 R^5 \nu_0(\lambda \lambda_0 + \mu \mu_0)}{r^3} \tag{5.137}$$

5.5.3 Comparison of the wobble potential with MacCullagh's formula

The MacCullagh formula for the gravitational potential U_G of a triaxial ellipsoid with mass E at an external point is given by (2.128), repeated here:

$$U_G = -G\frac{E}{r} - G\frac{A+B+C-3I}{2r^3} \tag{5.138}$$

I is the moment of inertia about a radial line passing through the point of observation. Substituting (5.106) for I with direction cosines $(\lambda_0, \mu_0, \nu_0)$ gives

$$
\begin{aligned}
U_G = & -G\frac{E}{r} \\
& -G\left(\frac{A+B+C-3\left(A\lambda_0^2 + B\mu_0^2 + C\nu_0^2 - 2K\lambda_0\mu_0 - 2H\mu_0\nu_0 - 2J\nu_0\lambda_0\right)}{2r^3}\right)
\end{aligned}
$$
$$\tag{5.139}$$

The terms involving products of inertia describe contributions to the potential from features that deviate from symmetry with respect to the x–y, y–z, and z–x planes. The potential of the deformation associated with the Chandler wobble depends on the products of direction cosines $\lambda_0\nu_0$ and $\mu_0\nu_0$. On comparing the coefficients of these products in (5.137) and (5.139) we get the following expressions for the products of inertia:

$$H = -\frac{\Omega^2 R^5 k}{3G}\mu \tag{5.140}$$

$$J = -\frac{\Omega^2 R^5 k}{3G}\lambda \tag{5.141}$$

5.5.4 Period of the Chandler wobble

The products of inertia H and J in the equations of motion (5.122) and (5.123) may now be replaced by the above expressions. The pair of simultaneous equations becomes

$$A\dot{\lambda} + \mu\Omega(C-A) - \frac{\Omega^3 R^5 k}{3G}\mu + \frac{\Omega^2 R^5 k}{3G}\dot{\lambda} = 0 \tag{5.142}$$

$$A\dot{\mu} - \lambda\Omega(C-A) + \frac{\Omega^3 R^5 k}{3G}\lambda + \frac{\Omega^2 R^5 k}{3G}\dot{\mu} = 0 \tag{5.143}$$

On regrouping the terms in these equations we get

$$\left(A + \frac{\Omega^2 R^5 k}{3G}\right)\dot{\lambda} + \mu\Omega\left((C - A) - \frac{\Omega^2 R^5 k}{3G}\right) = 0 \qquad (5.144)$$

$$\left(A + \frac{\Omega^2 R^5 k}{3G}\right)\dot{\mu} - \lambda\Omega\left((C - A) - \frac{\Omega^2 R^5 k}{3G}\right) = 0 \qquad (5.145)$$

Analogous equations (5.95) and (5.96) for the rigid Earth yielded the period of the free, Eulerian nutation,

$$\tau_0 = \frac{2\pi}{\Omega}\left(\frac{A}{C - A}\right) \qquad (5.146)$$

Proceeding in the same manner, the solutions of the nutation equations for an elastic Earth are reduced to a simple harmonic motion of the rotation axis with period

$$\tau = \frac{2\pi}{\Omega}\left(\frac{A + \Omega^2 R^5 k/(3G)}{C - A - \Omega^2 R^5 k/(3G)}\right) \qquad (5.147)$$

This is the period of the Chandler wobble. The numerator in (5.147) is larger than that in (5.146) and the denominator is smaller than that in (5.146). Thus the period of the Chandler wobble for the elastic Earth is longer than the period of the Eulerian nutation for a rigid Earth. The difference in periods can be used to compute a measure of the Earth's elastic yielding.

5.5.5 Calculation of Love's number *k* from the period of the Chandler wobble

Love's number k, which we encountered in the theory of the tides, is a measure of the global yielding of the Earth to the deforming tidal forces. A similar situation is encountered here: the elastic yielding of the Earth to the centrifugal force related to the free nutation results in the lengthened period observed in the Chandler wobble, which therefore depends on k.

The density distribution in the Earth is dependent on the ratio m between the centrifugal acceleration and the gravitational attraction at the equator (Box 3.2):

$$m = \frac{\omega^2 a}{GE/a^2} = \frac{\omega^2 a^3}{GE} \qquad (5.148)$$

Ignoring the small differences between the equatorial radius and mean radius, and using Ω for the Earth's rotation, we can replace this definition of m by

$$m \approx \frac{\Omega^2 R^3}{GE} \tag{5.149}$$

It follows that in (5.147) we can write

$$\frac{\Omega^2 R^5}{3G} = \frac{mER^2}{3} \tag{5.150}$$

$$\tau = \frac{2\pi}{\Omega} \left(\frac{A + kmER^2/3}{C - A - kmER^2/3} \right)$$
$$= \frac{2\pi}{\Omega} \left(\frac{A}{C - A} \right) \left(\frac{1 + kmER^2/(3A)}{1 - kmER^2/3(C - A)} \right) \tag{5.151}$$

$$\tau = \tau_0 \left(\frac{1 + kmER^2/(3A)}{1 - kmER^2/(3(C - A))} \right) \tag{5.152}$$

In (3.39) we established a relationship between the principal moments of inertia A and C, the flattening f, and the centrifugal acceleration ratio m,

$$\frac{C - A}{ER^2} = \frac{2f - m}{3} \tag{5.153}$$

and from (3.43) we know that the approximate values of A and C are

$$A \approx C \approx \frac{1}{3} ER^2 \tag{5.154}$$

We can substitute these values into (5.152), which simplifies to

$$\frac{1}{\tau} = \frac{1}{\tau_0} \left(1 - k \frac{m}{2f - m} \right) (1 + km)^{-1} \tag{5.155}$$

This relationship can be expanded as a binomial series. Neglecting second-order and higher powers of m and f, we obtain to first order

$$\frac{1}{\tau} = \frac{1}{\tau_0} \left(1 - k \frac{m}{2f - m} - km \right) = \frac{1}{\tau_0} \left(1 - km \left(\frac{1}{2f - m} - 1 \right) \right) \tag{5.156}$$

This reduces further to

$$\frac{1}{\tau} = \frac{1}{\tau_0} \left(1 - k \frac{m}{2f - m} \right) \tag{5.157}$$

By rearranging terms and solving for Love's number we get

$$k = \left(1 - \frac{\tau_0}{\tau}\right)\left(\frac{2f - m}{m}\right) \tag{5.158}$$

Upon inserting the known values for f, m, τ_0, and τ we get $k = 0.28$, in good agreement with the value obtained from the theory of the tides.

FURTHER READING

Lambeck, K. (1980). *The Earth's Variable Rotation: Geophysical Causes and Consequences.* Cambridge: Cambridge University Press, 464 pp.

Moritz, H. and Mueller, I. I. (1988). *Earth Rotation: Theory and Observation.* New York: Ungar, 617 pp.

Munk, W. H. and MacDonald, G. J. F. (1975). *The Rotation of the Earth: A Geophysical Discussion.* Cambridge: Cambridge University Press, 384 pp.

6

Earth's heat

The early thermal history of the Earth is a matter of some speculation. Current scientific consensus is that planet Earth formed by accretion of material with the same composition as chondritic meteorites. Accretion, a process that generated heat as colliding material gave up kinetic energy, led to differentiation of the planetary constituents into concentric layers. When the temperature of the early Earth reached the melting point of iron, the dense iron, accompanied by other siderophile elements such as nickel and sulfur, sank towards the center of the planet to form a liquid core. Meanwhile lighter elements rose to form an outer layer, the primitive mantle. Further differentiation took place later, creating a chemically different thin crust atop the mantle. Only the outer core is now molten, surrounding a solid inner core of iron that solidified out of the core fluid. Lighter elements left behind in the core rise through the core fluid and result in a composition-driven convection in the outer core, which is in addition to thermal convection. Although the short-term behavior of the mantle is like that of a solid, allowing the passage of seismic shear waves, its long-term behavior is characterized by plastic flow, so heat transport by convection or advection is possible. In the solid lithosphere and inner core heat is transported dominantly by thermal conduction.

The physical states of the Earth's mantle and core are well understood, but the variation of temperature with depth is not well known. Direct access is impossible and it is very difficult in laboratory experiments to achieve the temperatures and pressures in the Earth's deep interior. Consequently, some important thermodynamic parameters are inadequately known. Points on the melting-point curve can be determined from experiments at high temperature and pressure. Convection ensures that the temperature profile in the mantle and outer core is close to the adiabatic temperature curve, which can be calculated. From these considerations an approximate temperature profile in the Earth's interior can be estimated (Fig. 6.1). The temperatures in the mantle and outer core are close to the adiabatic curve, little temperature change occurs in the solid inner core, and comparatively rapid change occurs in the asthenosphere and lithosphere.

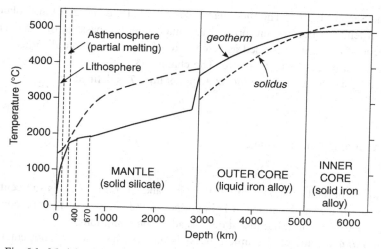

Fig. 6.1. Models of the adiabatic temperature profile (*geotherm*, solid curve) and the melting-point curve (*solidus*, dashed curve) in the Earth's interior. Data sources: tables in appendix G of Stacey and Davis (2008); for mantle solidus, Stacey (1992), appendix G.

6.1 Energy and entropy

Analysis of the thermal conditions in the Earth is based upon the First and Second Laws of Thermodynamics. The First Law is an application of the conservation of energy to a thermodynamic system. It states that energy cannot be created or destroyed in a closed system, but can only be transformed from one form to another. In an open system, extra terms must be considered to allow for the transfer of energy into or out the system (e.g., by the flow of matter). The total energy, Q, of a closed system consists of its internal energy, U, and the work, W, done in any external transfer of energy to the surroundings. The energy balance is expressed by the equation

$$dQ = dU + dW \tag{6.1}$$

Heat added to (or removed from) a closed system is used to increase the internal energy and to perform external work. For example, the gas molecules in a heated balloon are more energetic, and, if it is able to expand, the volume, V, increases. The external work dW due to the change in volume at constant pressure, P, is

$$dW = P\,dV \tag{6.2}$$

and so from the First Law of Thermodynamics the energy equation is

$$dQ = dU + P\,dV \tag{6.3}$$

The Second Law of Thermodynamics asserts that the energy of an isolated system tends to become uniformly distributed with the passage of time. The concept of *entropy*, *S*, is used as a measure of the microscopic disorder in a system at a particular temperature. The change dS in the entropy of a system caused by a change in energy dQ at a temperature T is defined as

$$dS = \frac{dQ}{T} \tag{6.4}$$

On substituting this into the energy equation we get

$$T\,dS = dU + P\,dV \tag{6.5}$$

This important relation, uniting the First and Second Laws, is the central equation of thermodynamics. It is important in the analysis of thermal conditions inside the Earth, because it defines adiabatic conditions.

An *adiabatic* thermodynamic process is one in which heat cannot enter or leave the system, i.e., $dQ = 0$. The entropy of an adiabatic reaction remains constant, because $dS = dQ/T = 0$. The adiabatic temperature gradient in the Earth serves as an important reference for estimates of the actual temperature gradient and for determining how heat is transferred.

6.2 Thermodynamic potentials and Maxwell's relations

The thermodynamic state of a system can be expressed with the aid of scalar functions called thermodynamic potentials. These are the internal energy, U, the enthalpy, H, the Helmholtz energy, A, and the Gibbs free energy, G. Each potential consists of a particular combination of the physical parameters pressure, temperature, volume, and entropy.

6.2.1 Thermodynamic potentials

Internal energy (U) has been described and defined above. A change in internal energy at constant temperature and pressure is related to changes in volume and entropy by

$$dU = T\,dS - P\,dV \tag{6.6}$$

Enthalpy (H) is a measure of the total energy of a system; it is a combination of the internal energy and the product of the pressure and volume:

$$H = U + PV \tag{6.7}$$

By taking the differentials of both sides of the equation we get

$$dH = dU + P\,dV + V\,dP \tag{6.8}$$

The conservation of energy, expressed in (6.5), allows us to reduce this to

$$dH = T\,dS + V\,dP \tag{6.9}$$

The **Helmholtz energy** (A) is defined from the relationship between the thermodynamic properties of macroscopic materials and their behavior on a microscopic level through statistical mechanics. It is a measure of the work obtainable from a closed thermodynamic system at constant temperature and constant volume, and is defined as

$$A = U - TS \tag{6.10}$$

Taking the differentials of both sides gives

$$dA = dU - T\,dS - S\,dT \tag{6.11}$$

Using (6.5), this becomes

$$dA = -P\,dV - S\,dT \tag{6.12}$$

The **Gibbs energy** (G) is defined in a similar way to the Helmholtz energy, but for constant pressure and temperature. It represents the maximum amount of energy obtainable from a closed system (i.e., one isolated from its surroundings) without increasing its volume, and is defined as

$$G = A + PV \tag{6.13}$$

The differentials give the equation

$$dG = dA + P\,dV + V\,dP \tag{6.14}$$

Combining this with (6.12) gives

$$dG = V\,dP - S\,dT \tag{6.15}$$

6.2.2 Maxwell's thermodynamic relations

Maxwell's relations are a set of partial differential equations derived from the definitions of the thermodynamic potentials that relate the parameters S, V, T, and P. The relations depend on the mathematical equality between the second derivatives of these parameters. This follows because the order of differentiation of a function $F(x, y)$ of two variables x and y is not important:

$$\frac{\partial}{\partial x}\left(\frac{\partial F}{\partial y}\right)_x = \frac{\partial^2 F}{\partial x\,\partial y} = \frac{\partial^2 F}{\partial y\,\partial x} = \frac{\partial}{\partial y}\left(\frac{\partial F}{\partial x}\right)_y$$

Maxwell's thermodynamic relations are derived in Box 6.1 by applying this condition to the different thermodynamic potentials. Summarized, they are

---

Box 6.1.   **Derivation of Maxwell's thermodynamic relations**

The **internal energy**, U, changes with V and S as in (6.6):

$$dU = T\,dS - P\,dV \tag{1}$$

dU can be written as a perfect differential using the partial derivatives of U with respect to V and S:

$$dU = \left(\frac{\partial U}{\partial S}\right)_V dS + \left(\frac{\partial U}{\partial V}\right)_S dV \tag{2}$$

The coefficients of dV and dS in these expressions must be equivalent, thus

$$P = -\left(\frac{\partial U}{\partial V}\right)_S \tag{3}$$

$$T = \left(\frac{\partial U}{\partial S}\right)_V \tag{4}$$

$$\frac{\partial P}{\partial S} = -\frac{\partial^2 U}{\partial T\,\partial S} \tag{5}$$

$$\frac{\partial T}{\partial V} = \frac{\partial^2 U}{\partial T\,\partial S} \tag{6}$$

$$\left(\frac{\partial T}{\partial V}\right)_S = -\left(\frac{\partial P}{\partial S}\right)_V \tag{7}$$

This is one of the Maxwell thermodynamic relations. The three others are obtained in a like manner.

The **enthalpy**, H, changes with P and T as in (6.9):

$$dH = T\,dS + V\,dP \tag{8}$$

dH can be written as a perfect differential using the partial derivatives of H with respect to T and P:

$$dH = \left(\frac{\partial H}{\partial S}\right)_P dS + \left(\frac{\partial H}{\partial P}\right)_S dP \qquad (9)$$

On equating the coefficients of dS and dP in these expressions, we have $T = (\partial H/\partial S)_P$ and $V = (\partial H/\partial P)_S$. Differentiating T with respect to P and V with respect to S gives

$$\left(\frac{\partial T}{\partial P}\right)_S = \left(\frac{\partial V}{\partial S}\right)_P \qquad (10)$$

The **Helmholtz energy**, A, changes with V and T as in (6.12):

$$dA = -P\, dV - S\, dT \qquad (11)$$

dA can be written as a perfect differential using the partial derivatives of A with respect to T and P:

$$dA = \left(\frac{\partial A}{\partial V}\right)_T dV + \left(\frac{\partial A}{\partial T}\right)_V dT \qquad (12)$$

On equating the coefficients of dV and dT in these expressions, we have $P = -(\partial A/\partial V)_T$ and $S = -(\partial A/\partial T)_V$. Differentiating P with respect to T and S with respect to V gives

$$\left(\frac{\partial P}{\partial T}\right)_V = \left(\frac{\partial S}{\partial V}\right)_T \qquad (13)$$

The **Gibbs energy**, G, changes with P and T as in (6.15):

$$dG = V\, dP - S\, dT \qquad (14)$$

dG can be written as a perfect differential using the partial derivatives of G with respect to T and P:

$$dG = \left(\frac{\partial G}{\partial P}\right)_T dP + \left(\frac{\partial G}{\partial T}\right)_P dT \qquad (15)$$

On equating the coefficients of dP and dT in these expressions, we have $V = (\partial G/\partial P)_T$ and $S = -(\partial G/\partial T)_P$. Differentiating V with respect to T and S with respect to P gives

$$\left(\frac{\partial V}{\partial T}\right)_P = -\left(\frac{\partial S}{\partial P}\right)_T \qquad (16)$$

$$\left(\frac{\partial T}{\partial V}\right)_S = -\left(\frac{\partial P}{\partial S}\right)_V \tag{6.16}$$

$$\left(\frac{\partial T}{\partial P}\right)_S = \left(\frac{\partial V}{\partial S}\right)_P \tag{6.17}$$

$$\left(\frac{\partial P}{\partial T}\right)_V = \left(\frac{\partial S}{\partial V}\right)_T \tag{6.18}$$

$$\left(\frac{\partial V}{\partial T}\right)_P = -\left(\frac{\partial S}{\partial P}\right)_T \tag{6.19}$$

6.3 The melting-temperature gradient in the core

The ambient pressure has a strong influence on the temperature at which the inner core solidifies from the core fluid. At the inner-core boundary the pressure is 330 GPa and the melting point of iron is around $T_m = 5,000$ K. If the latent heat of fusion of iron is L, the amount of heat exchanged when a mass m melts is $dQ = mL$ and the change in entropy is

$$dS = \frac{dQ}{T} = \frac{mL}{T_m} \tag{6.20}$$

Writing (6.17) in terms of full differentials, with $T = T_m$ and substituting (6.20) for dS, we have

$$\left(\frac{dT_m}{dP}\right)_S = \left(\frac{dV}{dS}\right)_P = \frac{V_L - V_S}{mL/T_m} \tag{6.21}$$

where V_L is the volume occupied by the mass of iron in a liquid state, and V_S is its volume in a solid state. We can write (6.21) as

$$\left(\frac{dT_m}{dP}\right)_S = \frac{T_m}{mL}(V_L - V_S) \tag{6.22}$$

This is known as the Clausius–Clapeyron equation for the change of state. During solidification the density changes from ρ_L for the liquid to ρ_S for the solid. The volume of a mass m of the material changes from $V_L = m/\rho_L$ before the change of state to $V_S = m/\rho_S$ after the change of state, so that

$$\frac{1}{T_m}\frac{dT_m}{dP} = \frac{1}{L}\left(\frac{1}{\rho_L} - \frac{1}{\rho_S}\right) \tag{6.23}$$

This equation must now be converted into a function of depth. The pressure inside the Earth is assumed to be hydrostatic. Under these conditions an increase in depth dz results in an increase in pressure dP solely because of the extra material added to the vertical column. If the local gravity at depth z is $g(z)$ and the local density is $\rho_L(z)$, the hydrostatic pressure increase is

$$dP = g(z)\rho_L(z)dz \tag{6.24}$$

On substituting this into (6.23), we get an equation relating the increase in melting temperature to increasing depth:

$$\frac{1}{T_m}\frac{dT_m}{dz} = \frac{g}{L}\left(1 - \frac{\rho_L}{\rho_S}\right) \tag{6.25}$$

The conditions in the core can be estimated from experiments and modeling. The melting temperature and the latent heat of fusion of iron at the enormous pressure in the core are not accurately known. For example, temperature estimates lie within the range 5,000–6,000 K. Some representative values of physical properties in the core are given in Table 6.1. Using values for the boundary between the inner and outer core in the modified Clausius–Clapeyron equation (6.25) the gradient of the melting temperature curve at that boundary is

$$\frac{dT_m}{dz} \approx 1.4 \text{ K km}^{-1} \tag{6.26}$$

Table 6.1. *Values of some physical parameters in the outer and inner core near to the core–mantle boundary (CMB) and inner-core boundary (ICB) (sources: (1) Dziewonski and Anderson, 1981; (2) Stacey, 2007)*

Physical property	Units	Outer core at CMB	Outer core at ICB	Inner core at ICB	Source
Gravity, g	m s^{-2}	10.7	4.4	4.4	1
Density, ρ	kg m^{-3}	9,900	12,160	12,980	1
Bulk modulus, K_S	GPa	646	1,300	1,300	1
$\Phi = K_S/\rho$	m^2 s^{-2}	67.3	107	107	1
Specific heat, c_P	J K^{-1} kg^{-1}	815	794	728	2
Temperature, T	K	3,700	5,000	5,000	2
Grüneisen parameter, γ		1.44	1.39	1.39	2
Volume expansion coefficient, α	10^{-6} K^{-1}	18.0	10.3	9.7	2
Latent heat of melting, L	10^5 J kg^{-1}	–	9.6	–	2

6.4 The adiabatic temperature gradient in the core

When heat is added to a material it causes an increase in temperature. The *specific heat* of the material is the amount of heat needed to raise the temperature of 1 kg of the material by 1 K; it can be defined for constant pressure, c_P, or constant volume, c_V. For a mass m of the material the heat dQ required to raise the temperature by dT at constant pressure is

$$dQ = mc_P\,dT \tag{6.27}$$

from which we get

$$\left(\frac{\partial Q}{\partial T}\right)_P = mc_P \tag{6.28}$$

The increase in temperature causes the material to expand. The coefficient of thermal expansion α_P is defined as the fractional increase in volume per degree increase in temperature. This can be written

$$\alpha_P = \frac{1}{V}\left(\frac{\partial V}{\partial T}\right)_P \tag{6.29}$$

The change in energy due to the heat added can be expressed as a perfect differential, giving

$$dQ = \left(\frac{\partial Q}{\partial T}\right)_P dT + \left(\frac{\partial Q}{\partial P}\right)_T dP \tag{6.30}$$

Using the definition of entropy, this becomes

$$T\,dS = \left(\frac{\partial Q}{\partial T}\right)_P dT + T\left(\frac{\partial S}{\partial P}\right)_T dP \tag{6.31}$$

Equation (6.28) can be used in the first term on the right, and the Maxwell relation from (6.19) can be used in the second term:

$$T\,dS = mc_P\,dT - T\left(\frac{\partial V}{\partial T}\right)_P dP \tag{6.32}$$

The condition for an adiabatic process, in which no heat is gained or lost by the system, is that the entropy remains constant, $dS = 0$, so

$$mc_P\,dT = TV\alpha_P\,dP \tag{6.33}$$

$$\left(\frac{\partial T}{\partial P}\right)_S = \frac{TV\alpha_P}{mc_P} = \frac{T\alpha_P}{\rho c_P} \tag{6.34}$$

This gives the adiabatic change of temperature with increasing pressure. Using (6.24), we convert the change in pressure to a change in depth and obtain the adiabatic temperature gradient,

$$\left(\frac{\partial T}{\partial z}\right)_S = \frac{gT\alpha_P}{c_P} \tag{6.35}$$

The depth profile of the adiabatic temperature is important for understanding conditions in the fluid core. If the actual temperature profile deviates from the adiabatic curve, this gives rise to convection currents, which redistribute the temperature to maintain adiabatic conditions. The physical parameters in Table 6.1 give an adiabatic temperature gradient in the fluid core of

$$\left(\frac{\partial T}{\partial z}\right)_S \approx 0.88 \text{ K km}^{-1} \tag{6.36}$$

at the core–mantle boundary, and

$$\left(\frac{\partial T}{\partial z}\right)_S \approx 0.29 \text{ K km}^{-1} \tag{6.37}$$

at the boundary with the inner core.

Comparison of these values with (6.26) shows that the melting temperature T_m increases more rapidly with depth than does the adiabatic temperature. In the early Earth, cooling from the surface, the melting temperature would have been reached first at the center. The core would have solidified from the bottom upwards, thus giving rise to the present layering of fluid outer core and solid inner core. Once the inner core became solid, it could cool further only by conduction, whereas convection continues to be the dominant process of heat transfer in the outer core.

6.5 The Grüneisen parameter

The atoms of a metal are located at specific sites in a regular lattice, forming a crystalline pattern that corresponds to the ambient conditions. Iron has a body-centered cubic (b.c.c.) structure at room pressure and temperature, but, as the pressure increases, the structure changes to a denser face-centered cubic (f.c.c.) packing, and eventually to hexagonal close packing (h.c.p.). At the pressure (330 GPa) and temperature (6,000 K) of the inner-core boundary iron is believed to have the h.c.p. structure. On a microscopic level the atoms in the iron lattice vibrate at a frequency given by the temperature. The atomic vibrations cannot take arbitrary values, but exhibit normal modes like classical vibrations of a string. The quantized vibrations, or *phonons*, are responsible for heat conduction in the solid and the long-wavelength phonons transport sound. A change in the temperature of a solid causes a change in volume, which alters the inter-atomic distances and thus the vibrational modes (phonon frequencies) of the crystal lattice. In solid-state physics this change is described by the Grüneisen parameter, γ. This is a dimensionless parameter, originally defined to represent the dependence of a particular mode of lattice vibration (phonon frequency) on a change of volume V. The *microscopic* definition of a Grüneisen parameter for a particular mode with frequency v_i is

$$\gamma_i = -\left(\frac{\partial \ln v_i}{\partial \ln V}\right)_T \tag{6.38}$$

It is difficult to adapt this definition to measurable quantities, because to do so requires detailed knowledge of the lattice dynamics. A more useful *macroscopic* definition of the Grüneisen parameter relates it to thermodynamic properties such as the bulk modulus, K_S, density, ρ, specific heat, c, and coefficient of thermal expansion, α. The definition at constant pressure is

$$\gamma = \frac{\alpha_P K_S}{\rho c_P} \tag{6.39}$$

The importance of γ in geophysics is due to its occurrence in equations that describe the dependence of physical properties on temperature and pressure, and therefore on depth. However, it is difficult to obtain values for the physical properties that define γ in laboratory experiments at high pressure and temperature that are representative for the conditions in the core. Conveniently, γ varies only slowly with pressure and temperature. It changes noticeably at Earth's important internal boundaries, but between these γ does not change much over large ranges of depth (Fig. 6.2).

Equation (6.35) for the adiabatic temperature gradient can be reformulated as follows:

$$\left(\frac{dT}{dz}\right)_S = \frac{g\rho T}{K_S}\left(\frac{\alpha_P K_S}{\rho c_P}\right) \tag{6.40}$$

Inserting the macroscopic definition of γ allows the temperature gradient to be written as

$$\left(\frac{dT}{dz}\right)_S = \gamma\frac{g\rho T}{K_S} \tag{6.41}$$

This equation can be refined further by using the velocities of seismic waves through the Earth, which are determined by the elastic constants. The relations

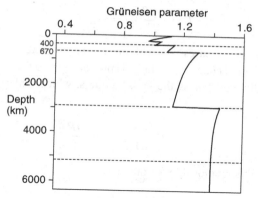

Fig. 6.2. Estimated variations of the Grüneisen parameter in different regions of the Earth's interior. Data source: Stacey and Davis (2008), appendix G.

between the P-wave velocity α and S-wave velocity β and the bulk modulus K_S, rigidity μ, and density ρ are developed in Section 8.5, giving

$$\alpha^2 = \frac{K_S + \frac{4}{3}\mu}{\rho} \qquad (6.42)$$

$$\beta^2 = \frac{\mu}{\rho} \qquad (6.43)$$

$$\frac{K_S}{\rho} = \alpha^2 - \frac{4}{3}\beta^2 = \Phi \qquad (6.44)$$

Φ is called the seismic parameter and is well known as a function of depth in the Earth because of the precise knowledge of seismic velocities on which Earth models such as PREM (Dziewonski and Anderson, 1981) are founded. Using this function, the equation for the adiabatic temperature gradient reduces to

$$\left(\frac{dT}{dz}\right)_S = \gamma \frac{gT}{\Phi} \qquad (6.45)$$

6.5.1 Temperature and density in the Earth

Thermal convection is the main form of heat transport in the outer core and is also important in the Earth's mantle. It keeps the ambient temperature close to the adiabatic temperature in these regions. Equation (6.41) for the adiabatic gradient can be reformulated as a function of pressure instead of depth,

$$dT = \gamma \frac{g\rho T}{K_S} dz = \gamma \frac{T}{K_S} dP \qquad (6.46)$$

When the pressure increases, the volume normally decreases. In an elastic material the fractional change in volume is proportional to the pressure change; the proportionality constant is the bulk modulus, which under adiabatic conditions is denoted K_S,

$$K_S = -V \left(\frac{dP}{dV}\right)_S = \rho \left(\frac{dP}{d\rho}\right)_S \qquad (6.47)$$

On rearranging this relationship we obtain

$$\frac{dP}{K_S} = \frac{d\rho}{\rho} \qquad (6.48)$$

Substituting into the adiabatic equation gives

$$\frac{dT}{T} = \gamma \frac{dP}{K_S} = \gamma \frac{d\rho}{\rho} \qquad (6.49)$$

Integrating both sides gives

$$\ln\left(\frac{T_2}{T_1}\right) = \gamma \ln\left(\frac{\rho_2}{\rho_1}\right) \qquad (6.50)$$

$$\frac{T_2}{T_1} = \left(\frac{\rho_2}{\rho_1}\right)^{\gamma} \qquad (6.51)$$

In this way, knowing the Grüneisen parameter for a particular domain allows the variation of temperature to be estimated from the variation of density with depth, which is well known.

6.6 Heat flow

When a straight conductor is heated so that one end is maintained at temperature T_1 and the other at a higher temperature T_2 (Fig. 6.3), the amount of heat ΔQ flowing out of the cooler end is inversely proportional to the length L of the conductor, and directly proportional to its cross-sectional area A, the measurement time Δt, and the temperature difference between the ends:

$$\Delta Q \propto A \frac{T_2 - T_1}{L} \Delta t \qquad (6.52)$$

We use this observation to define the vertical flow of heat at the Earth's surface.

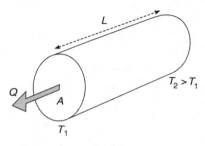

Fig. 6.3. The flow of heat Q along a conductor of length L and cross-section A, with ends maintained at different temperatures T_1 and T_2 ($T_2 > T_1$).

6.6.1 The heat-flow equation

Let Cartesian axes be defined so that the z-axis is vertically downwards and the x- and y-axes lie in the horizontal plane (Fig. 6.4). Consider the heat flowing vertically upwards along a very short conductor of cross-sectional area A_z normal to the z-direction and of length dz, such that its upper, cooler end at depth z has temperature T and the lower, warmer end at $z + dz$ has temperature $T + dT$. Upon inserting these values into (6.52) and introducing a proportionality constant k we obtain a differential equation for the heat loss per unit time:

$$\frac{dQ_z}{dt} = -kA_z\frac{dT}{dz} \tag{6.53}$$

The minus sign indicates that the heat flows in the direction of decreasing z (i.e., upwards). The proportionality constant is a material property of the conductor, namely its *thermal conductivity*. The heat flow q_z is defined as the heat crossing unit area per second:

$$q_z = \frac{1}{A_z}\frac{dQ_z}{dt} = -k\frac{dT}{dz} \tag{6.54}$$

This gives the vertical heat flow along the z-axis; it is possible to define horizontal components along the x- and y-axes in a similar way, so in general we can write the heat flow as a vector,

$$\mathbf{q} = -k\nabla T \tag{6.55}$$

6.6.2 The thermal-conduction equation

Returning to the one-dimensional situation, consider the heat flowing vertically upwards (along the z-axis) through a small rectangular box of sides Δx, Δy, and

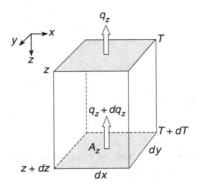

Fig. 6.4. Heat $Q_z + \Delta Q_z$ flows vertically into the base A_z of a small box with sides Δx, Δy, and Δz, whereas the amount of heat that leaves the top of the box is Q_z.

Δz with top surface at depth z, where the temperature is T (Fig. 6.4). The heat flow through the top surface is q_z, and the area of the surface normal to the flow is $A_z = \Delta x\,\Delta y$, so the total vertical loss of heat Q_z in time Δt is

$$Q_z = q_z(\Delta x \cdot \Delta y)\Delta t \qquad (6.56)$$

At depth $z + dz$ the heat entering the bottom end of the box is $Q_z + \Delta Q_z$, where

$$Q_z + \Delta Q_z = Q_z + \frac{\partial Q_z}{\partial z}\Delta z \qquad (6.57)$$

The amount of heat remaining in the box is the difference between the amounts entering and leaving it; on substituting from the right-hand side of (6.56) we have

$$\Delta Q_z = \frac{\partial Q_z}{\partial z}\Delta z = \frac{\partial q_z}{\partial z}\Delta z(\Delta x \cdot \Delta y)\Delta t \qquad (6.58)$$

Now we substitute the definition of the heat flow from (6.54) to obtain the amount of heat ΔQ_z retained in the box

$$\Delta Q_z = \frac{\partial}{\partial z}\left(-k\frac{\partial T}{\partial z}\right)\Delta V \Delta t = -k\frac{\partial^2 T}{\partial z^2}\Delta V \Delta t \qquad (6.59)$$

Let c_P be the specific heat at constant pressure and ρ the density of the material in the box, and let the rise in temperature caused by the extra heat be ΔT. The mass of matter in the box is $m = \rho\,\Delta V$, so, using the definition of specific heat,

$$\Delta Q_z = c_P m\,\Delta T = \rho c_P\,\Delta V \Delta T \qquad (6.60)$$

By equating this with (6.59) and deleting the factor ΔV on each side, we get

$$\rho c_P \frac{\partial T}{\partial t} = -k \frac{\partial^2 T}{\partial z^2} \tag{6.61}$$

$$\frac{\partial T}{\partial t} = -\left(\frac{k}{\rho c_P}\right) \frac{\partial^2 T}{\partial z^2} \tag{6.62}$$

The combination of physical parameters in parentheses defines the *thermal diffusivity*, κ,

$$\kappa = \frac{k}{\rho c_P} \tag{6.63}$$

The one-dimensional equation of heat conduction is therefore

$$\frac{\partial T}{\partial t} = -\kappa \frac{\partial^2 T}{\partial z^2} \tag{6.64}$$

This equation is one of the most important in geophysics. An equation with identical form describes the process of diffusion, by which a net flux of randomly moving particles that is proportional to the gradient in concentration of the particles can take place. Consequently, the *thermal-conduction equation* is sometimes called the *heat-diffusion equation*. Two specific examples of one-dimensional heat conduction are described in the following sections: the penetration of external heat into the Earth and the loss of heat from a cooling half-space.

By extension to the x- and y-directions, similar components are found, the only difference being that the second-order differentiation is with respect to x and y, respectively. The heat-conduction equation for three dimensions is

$$\frac{\partial T}{\partial t} = -\kappa \left(\frac{\partial^2 T}{\partial x^2} + \frac{\partial^2 T}{\partial y^2} + \frac{\partial^2 T}{\partial z^2}\right) \tag{6.65}$$

or

$$\frac{\partial T}{\partial t} = -\kappa \nabla^2 T \tag{6.66}$$

6.6.3 Penetration of solar heat in the Earth

Solar energy heats Earth's surface in a quasi-cyclical fashion, with a high and a low temperature each day, and a warmest and coldest month each year. The solar heat is transported downwards by conduction and is able to penetrate some distance into the Earth. The decay of temperature with depth below the surface can be evaluated by solving the one-dimensional heat-conduction equation with appropriate boundary conditions.

Let the z-axis again be the vertical direction. The temperature satisfying (6.64) is a function of both depth and time: $T = T(z, t)$. As in other cases, we apply the method of separation of variables. The depth variation is described by the function $Z(z)$ and the time variation by $\tau(t)$. Then

$$T(z, t) = Z(z)\tau(t) \tag{6.67}$$

This expression is inserted into the heat-conduction equation, and both sides are then divided by the product $Z(z)\tau(t)$. We have

$$Z\frac{\partial \tau}{\partial t} = \kappa\tau\frac{\partial^2 Z}{\partial z^2} \tag{6.68}$$

$$\frac{1}{\tau}\frac{\partial \tau}{\partial t} = \kappa\frac{1}{Z}\frac{\partial^2 Z}{\partial z^2} \tag{6.69}$$

Each side of this equation involves a different independent variable, thus both sides equal the same constant. This allows us to separate the equation into two parts. We must choose the constant to fit the boundary conditions of the stated problem. If the incident solar energy is a periodic function of time, then the solution will also be periodic. The time dependence of the surface temperature can be expressed by the real part of the complex function $\exp(i\omega t)$:

$$T = T_0 \cos(\omega t) = T_0 \operatorname{Re}(\exp(i\omega t)) \tag{6.70}$$

On comparing this with the left-hand side of (6.69), we see that the common constant in this equation must equal $i\omega$:

$$\frac{1}{\tau}\frac{\partial \tau}{\partial t} = i\omega \tag{6.71}$$

The time dependence of the temperature variation at depth is therefore

$$\tau = \tau_0 \exp(i\omega t) \tag{6.72}$$

Because both sides of (6.69) equal the same constant, the depth function satisfies

$$\kappa\frac{1}{Z}\frac{\partial^2 Z}{\partial z^2} = i\omega \tag{6.73}$$

$$\frac{\partial^2 Z}{\partial z^2} - i\frac{\omega}{\kappa}Z = 0 \tag{6.74}$$

This has the form of a simple harmonic equation,

$$\frac{\partial^2 Z}{\partial z^2} + n^2 Z = 0 \tag{6.75}$$

with solution

$$Z = Z_1 \exp(inz) + Z_0 \exp(-inz) \tag{6.76}$$

On comparing (6.74) and (6.75) we have

$$n^2 = -i\frac{\omega}{\kappa} \tag{6.77}$$

$$in = \sqrt{i\frac{\omega}{\kappa}} \tag{6.78}$$

As shown in Section 1.2, the complex number $\exp(i\theta)$ can be written

$$\exp(i\theta) = \cos\theta + i\sin\theta \tag{6.79}$$

Thus

$$i = \exp\left(i\frac{\pi}{2}\right) \tag{6.80}$$

and

$$\sqrt{i} = \exp\left(i\frac{\pi}{4}\right) = \cos\left(\frac{\pi}{4}\right) + i\sin\left(\frac{\pi}{4}\right) = \frac{1}{\sqrt{2}}(1+i) \tag{6.81}$$

Equation (6.78) can now be written

$$in = \sqrt{\frac{\omega}{2\kappa}}(1+i) \tag{6.82}$$

Upon inserting this into (6.76), the variation of temperature with depth becomes

$$Z = Z_1 \exp\left(\sqrt{\frac{\omega}{2\kappa}}(1+i)z\right) + Z_0 \exp\left(-\sqrt{\frac{\omega}{2\kappa}}(1+i)z\right) \tag{6.83}$$

In this problem of solar heating we are interested in the flow of heat downwards into the Earth, in the $+z$-direction. The temperature fluctuation related to solar heating decreases with increasing depth, thus dZ/dz must be negative. The first term in (6.83) increases exponentially with depth, so we exclude it by setting $Z_1 = 0$ and obtain

$$T(z,t) = Z_0 \exp\left(-\sqrt{\frac{\omega}{2\kappa}}(1+i)z\right) \cdot \tau_0 \exp(i\omega t) \tag{6.84}$$

The initial conditions at the surface (depth $z = 0$, time $t = 0$) are that the temperature is equal to T_0. Thus $Z_0\tau_0 = T_0$ and the solution to the heat-conduction equation is

$$T(z, t) = T_0 \exp\left(-\sqrt{\frac{\omega}{2\kappa}}z\right) \exp\left\{i\left(\omega t - \sqrt{\frac{\omega}{2\kappa}}z\right)\right\} \qquad (6.85)$$

The temperature variation with time and depth is the real part of this solution:

$$T(z, t) = T_0 \exp\left(-\frac{z}{d}\right) \cos\left(\omega t - \frac{z}{d}\right) \qquad (6.86)$$

We have simplified the result by using

$$d = \sqrt{\frac{2\kappa}{\omega}} \qquad (6.87)$$

This is a characteristic depth for the problem, often called the penetration depth. It is the depth at which the temperature fluctuation has decreased to $1/e$ of its surface value. It depends both on the frequency of the fluctuation and on the material properties of the ground. The thermal diffusivity is defined on the basis of the specific heat, density, and thermal conductivity, all of which vary with temperature. Consequently the thermal diffusivity is temperature-dependent; in common rocks it decreases with increasing temperature. Assuming representative values of the physical properties of some common near-surface rock types, typical penetration depths can be calculated (Table 6.2). The penetration depth of the daily temperature variation (period = 86,400 s, $\omega = 7.27 \times 10^{-5}$ rad s^{-1}) is around 18 cm; that of the annual fluctuation (period = 3.15×10^7 s, $\omega = 1.99 \times 10^{-7}$ rad s^{-1}) is around 3.5 m.

Table 6.2. *Calculated penetration depths of solar energy in continental surface rocks for daily and annual temperature fluctuations (source: average values from graphed data in Vosteen and Schellschmidt (2003))*

Thermal property	Units	Mean value
Thermal conductivity, k	W m^{-1} K^{-1}	2.5
Specific heat, c_P	J kg^{-1} K^{-1}	800
Density, ρ	kg m^{-3}	2,750
Thermal diffusivity, κ	10^{-6} m^2 s^{-1}	1.1
Penetration depth of daily fluctuation	m	0.18
Penetration depth of annual fluctuation	m	3.4

Note that the penetration depth d is not the maximum depth to which the solar energy can penetrate, but merely the depth at which the amplitude sinks to $1/e$. The surface temperature change is felt well below the penetration depth. At a depth of $5d$ the signal has attenuated to about 1% of the surface value.

The attenuation of the surface temperature fluctuation is accompanied by a shift in phase of the signal. We can write (6.86) as

$$T(z, t) = T_0 \exp\left(-\frac{z}{d}\right) \cos(\omega(t - t_0)) \tag{6.88}$$

The time t_0 represents a delay in the time at which the surface extreme values are felt at depth z:

$$t_0 = \frac{z}{\omega d} = \frac{z}{\omega}\sqrt{\left(\frac{\omega}{2\kappa}\right)} = \frac{z}{\sqrt{2\kappa\omega}} \tag{6.89}$$

Figure 6.5 shows the attenuation and phase shift of the temperature for a hypothetical sedimentary rock, using the data in Table 6.2. The surface temperature is assumed to vary periodically between +10 °C and −10 °C. At depths below about 1 m the daily surface change is barely discernible; the corresponding depth for the annual fluctuation is about 19 m. At depth $z = \pi d$ (around 11 m in this case) the phase shift of the annual variation with respect to surface values is 180°; i.e., when the surface temperature is at its peak, the temperature at this depth is minimum.

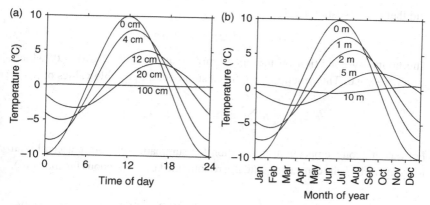

Fig. 6.5. Effect of surface solar heating on near-surface temperatures in a sedimentary rock. Attenuation and phase shift of (a) daily and (b) annual temperature fluctuations.

6.6.4 Cooling of a semi-infinite half-space

The second application of the heat-conduction equation is to the outward vertical flow of heat from the Earth's interior as it cools from an initially hot state. We assume a one-dimensional model consisting of a semi-infinite half-space that extends to infinity in the (vertical) z-direction. Lateral components of heat flow, such as result from modification by surface topography, are ignored. The problem consists of determining the temperature distribution $T(z, t)$ as a function of depth z in the half-space at time t after it starts to cool.

Let the temperature of the upper surface be zero. The temperature in the cooling half-space must satisfy the heat-conduction equation, and is obtained by separation of the variables as in (6.69):

$$\frac{1}{\kappa}\frac{1}{\tau}\frac{\partial \tau}{\partial t} = \frac{1}{Z}\frac{\partial^2 Z}{\partial z^2} \tag{6.90}$$

In this instance we are studying not a fluctuating temperature, but a steady cooling process. Separating the variables as before, we set the separation constant equal to $-n^2$:

$$\frac{1}{\kappa}\frac{1}{\tau}\frac{\partial \tau}{\partial t} = -n^2 \tag{6.91}$$

$$\frac{1}{Z}\frac{\partial^2 Z}{\partial z^2} = -n^2 \tag{6.92}$$

The particular solution of the time-dependent part is

$$\tau = \tau_0 \exp(-\kappa n^2 t) \tag{6.93}$$

and that of the spatial part is

$$Z = A_n \cos(nz) + B_n \sin(nz) \tag{6.94}$$

The boundary condition on the upper surface at $z = 0$ is $T(0, t) = 0$, which requires $A_n = 0$. The general solution is a sum over all possible values of n:

$$T(z, t) = \tau_0 \sum_{n=0}^{\infty} \exp(-\kappa n^2 t) B_n \sin(nz) \tag{6.95}$$

For a continuous temperature distribution the summation can be replaced by an integral in which the constants τ_0 and B_n are combined in a continuous function $B(n)$:

$$T(z, t) = \int_{n=0}^{\infty} \exp(-\kappa n^2 t) B(n) \sin(nz) dn \tag{6.96}$$

Suppose that at $t = 0$ the cooling half-space has an initial temperature distribution $T(z)$:

$$T(z, 0) = T(z) = \int_{n=0}^{\infty} B(n)\sin(nz)dn \qquad (6.97)$$

This is a Fourier integral equation, in which the amplitude function $B(n)$ must be determined. This is obtained by using the properties of Fourier sine transforms, which are explained briefly in Section 1.17. The Fourier sine transform allows us to write the amplitude function as

$$B(n) = \frac{2}{\pi} \int_{z=0}^{\infty} T(z)\sin(nz)dz = \frac{2}{\pi} \int_{\zeta=0}^{\infty} T(\zeta)\sin(n\zeta)d\zeta \qquad (6.98)$$

In the final expression the integration variable has been changed from z to ζ to avoid subsequent confusion when we insert the result back into (6.96). The substitution gives

$$T(z, t) = \frac{2}{\pi} \int_{\zeta=0}^{\infty} T(\zeta) \left[\int_{n=0}^{\infty} \exp(-\kappa n^2 t)\sin(nz)\sin(n\zeta)dn \right] d\zeta \qquad (6.99)$$

Now we can change the integrand by using the trigonometric relationship

$$2\sin(nz)\sin(n\zeta) = \cos(n(\zeta - z)) - \cos(n(\zeta + z)) \qquad (6.100)$$

giving

$$T(z, t) = \frac{1}{\pi} \int_{\zeta=0}^{\infty} T(\zeta) \left[\int_{n=0}^{\infty} (\exp(-\kappa n^2 t)\cos(n(\zeta - z)) \right.$$

$$\left. - \exp(-\kappa n^2 t)\cos(n(\zeta + z)))dn \right] d\zeta \qquad (6.101)$$

Each of the integrals inside the square brackets has the same form, namely $\int_{n=0}^{\infty} \exp(-an^2)\cos(nu)dn$, with $a = \kappa t$ and $u = \zeta - z$ or $u = \zeta + z$, respectively. The integration of this function is shown in Box 6.2 to be

$$\int_{n=0}^{\infty} \exp(-an^2)\cos(nu)dn = \frac{1}{2}\sqrt{\frac{\pi}{a}}\exp\left(-\frac{u^2}{4a}\right) \qquad (6.102)$$

Applying this solution to each integral in the square brackets in (6.101), with $a = \kappa t$, gives

Box 6.2. **The cooling half-space integration**

The cooling half-space solution requires evaluation of the integral

$$Y = \int_{n=0}^{n=\infty} \exp(-an^2)\cos(nu)\,dn \tag{1}$$

Note that, on differentiating with respect to u,

$$\frac{\partial Y}{\partial u} = \int_{n=0}^{n=\infty} -n\exp(-an^2)\sin(nu)\,dn \tag{2}$$

Integrating (2) by parts with respect to n gives

$$\frac{\partial Y}{\partial u} = \left[\frac{\exp(-an^2)}{2\alpha}\sin(nu)\right]_0^{\infty} - \frac{u}{2\alpha}\int_0^{\infty}\exp(-an^2)\cos(nu)\,dn = -\frac{u}{2\alpha}Y \tag{3}$$

$$\frac{1}{Y}\frac{\partial Y}{\partial u} = \frac{\partial}{\partial u}\ln(Y) = -\frac{u}{2\alpha} \tag{4}$$

$$\ln(Y) = -\frac{u^2}{4\alpha} + \ln(Y_0)$$

Here we have introduced Y_0 as a constant of integration, and the solution to the integration is

$$Y = Y_0\exp\left(-\frac{u^2}{4\alpha}\right) \tag{5}$$

The constant Y_0 is the value of the integral Y for $u = 0$. This constant may be determined as follows:

$$Y_0 = \int_{x=0}^{\infty}\exp(-ax^2)\,dx = \int_{y=0}^{\infty}\exp(-ay^2)\,dy \tag{6}$$

$$(Y_0)^2 = \left(\int_{x=0}^{\infty}\exp(-ax^2)\,dx\right)\left(\int_{y=0}^{\infty}\exp(-ay^2)\,dy\right)$$

$$= \int_{x=0}^{\infty}\int_{y=0}^{\infty}\exp(-a(x^2+y^2))\,dx\,dy \tag{7}$$

On changing to polar coordinates (r, θ), we have $x = r\cos\theta$ and $y = r\sin\theta$, and the element of area becomes $dx\,dy = r\,dr\,d\theta$. The limits of integration change from $(0 \le x \le \infty; 0 \le y \le \infty)$ to $(0 \le r \le \infty; 0 \le \theta \le \pi/2)$:

$$(Y_0)^2 = \int_{\theta=0}^{\pi/2} \int_{r=0}^{\infty} \exp(-\alpha r^2)r\,dr\,d\theta = \int_{\theta=0}^{\pi/2} \left(\int_{r=0}^{\infty} \exp(-\alpha r^2)r\,dr \right) d\theta \quad (8)$$

$$(Y_0)^2 = \int_{\theta=0}^{\pi/2} \left[-\frac{\exp(-\alpha r^2)}{2\alpha} \right]_{r=0}^{\infty} d\theta = \frac{1}{2\alpha} \int_{\theta=0}^{\pi/2} d\theta = \frac{\pi}{4\alpha} \quad (9)$$

$$Y_0 = \frac{1}{2}\sqrt{\frac{\pi}{\alpha}} \quad (10)$$

By inserting this value into (5) we get the evaluated integral:

$$Y = \frac{1}{2}\sqrt{\frac{\pi}{\alpha}} \exp\left(-\frac{u^2}{4\alpha}\right) \quad (11)$$

$$T(z, t) = \frac{1}{2\sqrt{\pi\kappa t}} \int_{\zeta=0}^{\infty} T(\zeta)\left[\exp\left(-\frac{(\zeta-z)^2}{4\kappa t}\right) - \exp\left(-\frac{(\zeta+z)^2}{4\kappa t}\right) \right] d\zeta$$

$$(6.103)$$

If the cooling body has initially a *uniform* temperature T_0, then $T(z) = T_0$ and the temperature distribution can be written

$$T(z, t) = \frac{T_0}{2\sqrt{\pi\kappa t}} \left\{ \int_{\zeta=0}^{\infty} \exp\left(-\frac{(\zeta-z)^2}{4\kappa t}\right) d\zeta - \int_{\zeta=0}^{\infty} \exp\left(-\frac{(\zeta+z)^2}{4\kappa t}\right) d\zeta \right\}$$

$$(6.104)$$

In the first integration, on writing $w = (\zeta - z)/(2\sqrt{\kappa t})$, we have $dw = [1/(2\sqrt{\kappa t})]d\zeta$ and the upper and lower limits of the integration change to ∞ and $-z/(2\sqrt{\kappa t})$, respectively. Similarly, on writing $v = (\zeta + z)/(2\sqrt{\kappa t})$ in the second integration, we get an equivalent expression for dv, but the integration limits become ∞ and $z/(2\sqrt{\kappa t})$, respectively. Equation (6.104) becomes

$$T(z, t) = \frac{T_0}{\sqrt{\pi}} \left\{ \int_{w=-z/(2\sqrt{\kappa t})}^{\infty} \exp(-w^2)dw - \int_{v=z/(2\sqrt{\kappa t})}^{\infty} \exp(-v^2)dv \right\}$$

$$(6.105)$$

The integration variables w and v in this equation are interchangeable, and can be combined in a single integration, modifying the integration limits accordingly. This gives

$$T(z, t) = \frac{T_0}{\sqrt{\pi}} \left\{ \int_{w=-z/(2\sqrt{\kappa t})}^{z/(2\sqrt{\kappa t})} \exp(-w^2)dw \right\} = \frac{2T_0}{\sqrt{\pi}} \left\{ \int_{w=0}^{z/(2\sqrt{\kappa t})} \exp(-w^2)dw \right\}$$

$$(6.106)$$

$$T(z, t) = T_0 \left\{ \frac{2}{\sqrt{\pi}} \int_{w=0}^{z/(2\sqrt{\kappa t})} \exp(-w^2)dw \right\} \qquad (6.107)$$

The expression in brackets is the *error function* (Box 6.3), defined as

Box 6.3.   **The error function**

The *error function* is closely related to the bell-shaped normal distribution. However, only positive values of the independent variable u are considered, so the graph of the defining function is similar to the right half of a normal distribution as in Fig. B6.3(a). Its equation is

$$f(u) = \frac{2}{\sqrt{\pi}} \exp(-u^2) \qquad (1)$$

The error function erf(η) is defined as the area under this curve from the origin at $u = 0$ to the value $u = \eta$:

$$\text{erf}(\eta) = \frac{2}{\sqrt{\pi}} \int_0^\eta \exp(-u^2)du \qquad (2)$$

The complementary error function, erfc(η), is defined as

$$\text{erfc}(\eta) = 1 - \text{erf}(\eta) = \frac{2}{\sqrt{\pi}} \int_\eta^\infty \exp(-u^2)du \qquad (3)$$

The value of erf(η) or erfc(η) for any particular value of η may be obtained from standard tables, or from a graph like Fig. B6.3(b).

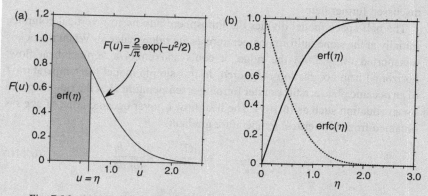

Fig. B6.3. (a) The error function erf(η) is defined as the area under the normal distribution curve from the origin at $u = 0$ to the value $u = \eta$. (b) Graphs of the error function erf(η) and complementary error function erfc(η).

$$\mathrm{erf}(\eta) = \frac{2}{\sqrt{\pi}} \int_{u=0}^{\eta} \exp\left(-u^2\right) du \tag{6.108}$$

Values of the error function are tabulated for any finite argument. The solution for the temperature distribution as a function of time and depth in the cooling half-space is therefore

$$T(z, t) = T_0 \, \mathrm{erf}\left(\frac{z}{2\sqrt{\kappa t}}\right) \tag{6.109}$$

This equation allows us to understand the heat flow measured over oceanic crust.

6.6.5 Cooling of oceanic lithosphere

In plate-tectonic theory the oceanic lithosphere is formed at a ridge axis and is transported away from the ridge by sea-floor spreading, cooling as it does so. The age, or cooling-time t, of the lithosphere at any place is proportional to its distance from the ridge axis, assuming a constant spreading rate. Two models are in common use: a one-dimensional half-space model as described above,

and a plate model that considers the lithosphere to be a cooling boundary layer
with its top surface at sea-floor temperature, and with its base and the edge at the
spreading ridge at the temperature of the asthenosphere. The first of these is
discussed further here.

The half-space model divides the lithosphere into narrow vertical columns,
initially at the same uniform temperature as the ridge material. When a block is
transported away from the ridge, it cools and emits a vertical heat flow;
horizontal heat conduction is ignored. In this simple model the temperature T
of an oceanic plate at a time t after forming at temperature T_0 at the ridge is given
by an equation such as (6.109). The heat flow q_z over oceanic crust of age t is
obtained from the vertical temperature gradient:

$$q_z = -k\frac{dT}{dz} = -k\frac{dT}{d\eta}\frac{d\eta}{dz} \tag{6.110}$$

$$\frac{dT}{d\eta} = T_0\frac{d}{d\eta}\operatorname{erf}(\eta) = \frac{2T_0}{\sqrt{\pi}}\frac{d}{d\eta}\int_{u=0}^{\eta}\exp(-u^2)\,du$$
$$= \frac{2T_0}{\sqrt{\pi}}\exp(-\eta^2) \tag{6.111}$$

$$\frac{d\eta}{dz} = \frac{d}{dz}\left(\frac{z}{2\sqrt{\kappa t}}\right) = \frac{1}{2\sqrt{\kappa t}} \tag{6.112}$$

On combining these equations, we obtain the heat flow

$$q_z = \frac{T_0}{\sqrt{\pi\kappa t}}\exp(-\eta^2) \tag{6.113}$$

At the surface of the oceanic plate, $\eta = z = 0$, and $\exp(-\eta^2) = 1$, so the heat flow
over crust of age t is given by

$$q_z = \frac{T_0}{\sqrt{\pi\kappa t}} \tag{6.114}$$

The inverse-square-root dependence on age predicted by the half-space
model agrees well with observed oceanic heat-flow values (Fig. 6.6).

Heat-flow data in young sea floor, and where sediment cover is thin, are
systematically biased by hydrothermal circulation, which transports some of the
heat by advection. This can be compensated for by considering only sites that
have sufficient sediment cover and are far enough from basement outcrops that
hydrothermal circulation perturbations are minimal. In particular, Fig. 6.6
shows sites on young sea floor where detailed investigations (seismic imaging

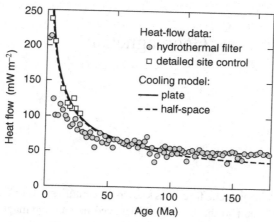

Fig. 6.6. Oceanic heat-flow data from all the oceans, plotted versus lithospheric age. The data have been filtered to exclude sites where sediment thickness is less than 325 m and those which are within 85 km of a seamount. Solid dots show median heat flow for 2-Myr age bins; open squares represent high-quality data from sites where the environment of the site is known from seismic imaging of the sea floor and other geophysical investigations. The dashed and solid lines represent heat flow for the half-space and plate cooling models, respectively. After Hasterok (2010).

of the buried basement topography, closely spaced heat-flow measurements and profiles) have been carried out. The heat-flow values at these sites agree very well with the predictions of both cooling models. For older oceanic lithosphere the plate model fits the data more closely than the half-space model and appears to be a better overall model.

FURTHER READING

Anderson, O. L. (2007). Grüneisen's parameter for iron and Earth's core, in *Encyclopedia of Geomagnetism and Paleomagnetism*, ed. D. Gubbins and E. Herrero-Bervera. Dordrecht: Springer, pp. 366–373.

Carslaw, H. S. and Jaeger, J. C. (2001). *Conduction of Heat in Solids*. Oxford: Clarendon Press, 510 pp.

Jessop, A. M. (1990). *Thermal Geophysics*. Amsterdam: Elsevier, 306 pp.

Özişik, M. N. (1980). *Heat Conduction*. New York: John Wiley & Sons, 687 pp.

Stein, C. A. (1995). Heat flow of the Earth, in *Global Earth Physics: A Handbook of Physical Constants*, ed. T. J. Ahrens. Washington, DC: American Geophysical Union, pp. 144–158.

7

Geomagnetism

The existence of a magnetic force was known for centuries before William Gilbert pointed out in 1600 that the Earth itself behaved like a huge magnet. Gradually maps were made of the geomagnetic elements. Systematic investigation of magnetic behavior was undertaken in the late eighteenth and early nineteenth centuries. The French scientist Charles Augustin de Coulomb showed experimentally that forces of attraction and repulsion exist between the ends of long thin magnetized rods, and that they obey rules similar to those determining the interaction of electrical charges. A freely suspended magnet was observed to align approximately north–south; the north-seeking end became known as its north pole, the opposite end as its south pole. The origin of magnetic force was attributed to magnetic charges, which, through association, became known as magnetic poles. Subsequently, it was shown that individual magnetic poles, or monopoles, do not exist. All magnetic fields originate in electric currents. This is true even at atomic dimensions; circulating (and spinning) electrical charges impart magnetic properties to atoms. However, the concept of multiple pole combinations (e.g., the dipole, quadrupole, and octupole) proved to be very useful for describing the geometries of magnetic fields.

7.1 The dipole magnetic field and potential

The most important field geometry is that of a magnetic dipole. This was originally imagined to consist of two equal and opposite magnetic poles that lie infinitesimally close to each other (Appendix A2). At distances several times greater than the size of the source the field of a very short bar magnet is very nearly a dipole field, as is the magnetic field produced by an electric current in a small plane loop. In an external magnetic field $\mathbf{B}$ a magnetic dipole experiences a torque τ that aligns it with the field (Appendix A4). The torque is governed by the relationship

$$\mathbf{\tau} = \mathbf{m} \times \mathbf{B} \tag{7.1}$$

In this equation $\mathbf{m}$ is the *magnetic moment* of the dipole, a measure of its strength. For a current-carrying loop $\mathbf{m}$ is equal to the product of the current I in the loop and its area A, and its direction $\mathbf{e_n}$ is that of the normal to the plane of the loop (Appendix A4):

$$\mathbf{m} = (IA)\mathbf{e_n} \tag{7.2}$$

The dimensions of magnetic moment are by this definition A m^2; the dimensions of torque are N m; thus the SI unit of the magnetic field $\mathbf{B}$, the *tesla*, has the dimensions N A^{-1} m^{-1}.

The potential W of a dipole magnetic moment $\mathbf{m}$ at distance r from its center and at an azimuthal angle θ between the dipole axis and the radial direction is (Appendix A2)

$$W = \frac{\mu_0}{4\pi} \frac{m \cos \theta}{r^2} \tag{7.3}$$

The constant μ_0 is the magnetic field constant. It is defined in SI units to be exactly $4\pi \times 10^{-7}$ N A^{-2} (alternatively designated henry m^{-1}). The dipole potential is the most important component of the geomagnetic field, representing more than 93% of its energy density.

The dipole magnetic field $\mathbf{B}$ is the gradient of the dipole potential: $\mathbf{B} = -\nabla W$. In spherical coordinates the field has a radial component B_r and an azimuthal component B_θ. These are

$$B_r = -\frac{\partial}{\partial r}\left(\frac{\mu_0}{4\pi}\frac{m \cos \theta}{r^2}\right) = \frac{\mu_0}{4\pi}\frac{2m \cos \theta}{r^3} \tag{7.4}$$

$$B_\theta = -\frac{1}{r}\frac{\partial}{\partial \theta}\left(\frac{\mu_0}{4\pi}\frac{m \cos \theta}{r^2}\right) = \frac{\mu_0}{4\pi}\frac{m \sin \theta}{r^3} \tag{7.5}$$

For a dipole at the center of the spherical Earth, the azimuthal component of the field, B_θ, is horizontal. Moreover, if the dipole is aligned with the Earth's axis, the angle θ is the complement of the magnetic latitude β. The direction of the field makes an angle I with the horizontal called the inclination of the field (see Fig. 7.1(b) and Appendix A, Fig. A1). The inclination, magnetic co-latitude, and magnetic latitude are related by

$$\tan I = \frac{B_r}{B_\theta} = 2 \cot \theta = 2 \tan \beta \tag{7.6}$$

This equation forms the basis of paleomagnetic determination of ancient paleolatitudes from the inclinations of remanent magnetizations measured in oriented rock samples.

7.2 Potential of the geomagnetic field

The empirical laws that govern electricity and magnetism are summarized in Maxwell's equations (Appendix B). Analysis of the present geomagnetic field requires Gauss's law and Ampère's law.

Gauss's law established that the net magnetic flux through any closed surface is zero. This is equivalent to stating that there are no magnetic monopoles: dipole sources such as current circuits, even at atomic scale, produce zero net flux through a surrounding surface. The corresponding equation is

$$\nabla \cdot \mathbf{B} = 0 \tag{7.7}$$

Ampère's law showed that an electric current produces a magnetic field in the surrounding space, and it relates the strength of the magnetic field $\mathbf{B}$ to the electric field $\mathbf{E}$ that causes the current:

$$\nabla \times \mathbf{B} = \mu_0 \sigma \mathbf{E} + \mu_0 \varepsilon_0 \frac{\partial \mathbf{E}}{\partial t} \tag{7.8}$$

The first term on the right is the electric current associated with the flow of free charges in a conductor and relies on Ohm's law; the second term is the electric displacement current that results from time-dependent motions of charges bound to a parent atom. The parameter μ_0 is the magnetic field constant, or *permeability* of free space, and ε_0 is the electric field constant, or *permittivity* of free space; σ is the electrical conductivity of the medium.

In a region that is free of sources of the magnetic field (such as the space just above the Earth's surface in which the field is measured), we can assume that there are no electric or displacement currents, thus

$$\nabla \times \mathbf{B} = 0 \tag{7.9}$$

Consequently, the magnetic field $\mathbf{B}$ can be written as the gradient of a scalar potential, W:

$$\mathbf{B} = -\nabla W \tag{7.10}$$

On substituting for $\mathbf{B}$ in (7.7) the potential W of the Earth's magnetic field is seen to satisfy Laplace's equation:

$$\nabla^2 W = 0 \tag{7.11}$$

7.2.1 The fields of internal and external origin

The geomagnetic potential at Earth's surface arises from two sources. The most important part of the field originates in the Earth's interior, and the rest originates outside the Earth, e.g., from current systems in the ionosphere. Let W_i be the potential of the field of internal origin and W_e be the potential of the field of external origin. The total geomagnetic potential W at Earth's surface is

$$W = W_e + W_i \qquad (7.12)$$

The geomagnetic potential has to be conformable with Earth's approximately spherical geometry, so the solution of (7.11) requires spherical polar coordinates. The general solution of Laplace's equation is therefore as described in Section 1.16. The variation of potential on a spherical surface is described by spherical harmonic functions of the co-latitude θ and longitude ϕ. The variation of potential with radial distance r consists of two parts. In a region where r can be zero, the potential is proportional to r^n. At Earth's surface this condition applies to the field due to sources outside the Earth, so W_e must vary as r^n. In a region where r can be very large or infinite, the potential is proportional to $1/r^{n+1}$. Outside the Earth and on its surface, this applies to the potential of the field of internal origin, so W_i must vary as $1/r^{n+1}$. These considerations lead to the following definition for the potential W_e of the field of external origin:

$$W_e = R \sum_{n=1}^{\infty} \sum_{m=0}^{n} \left(\frac{r}{R}\right)^n \left(G_n^m \cos(m\phi) + H_n^m \sin(m\phi)\right) P_n^m (\cos\theta), \ r < R$$

$$(7.13)$$

Similarly, the potential W_i of the field of internal origin is

$$W_i = R \sum_{n=1}^{\infty} \sum_{m=0}^{n} \left(\frac{R}{r}\right)^{n+1} \left(g_n^m \cos(m\phi) + h_n^m \sin(m\phi)\right) P_n^m (\cos\theta), \quad r > R$$

$$(7.14)$$

Terms with $n = 0$ are absent from these expressions because magnetic monopoles do not exist. At the Earth's surface the expressions simplify to

$$W_e = R \sum_{n=1}^{\infty} \sum_{m=0}^{n} \left(G_n^m \cos(m\phi) + H_n^m \sin(m\phi)\right) P_n^m (\cos\theta) \qquad (7.15)$$

$$W_i = R \sum_{n=1}^{\infty} \sum_{m=0}^{n} \left(g_n^m \cos(m\phi) + h_n^m \sin(m\phi)\right) P_n^m (\cos\theta) \qquad (7.16)$$

In a convention adopted in 1939 by the scientific body that preceded the modern International Association of Geomagnetism and Aeronomy (IAGA), it was agreed to base the spherical harmonic functions in the magnetic potential on the partially normalized Schmidt polynomials (Section 1.15.2). The coefficients

(g_n^m, h_n^m) and (G_n^m, H_n^m) are called the Gauss (or Gauss–Schmidt) coefficients of the fields of internal and external origin, respectively. They have the dimensions of magnetic field and their magnitudes diagnose the relative importance of the external and internal sources of the field.

7.2.2 Determination of the Gauss coefficients

It is not possible to measure the geomagnetic potential directly, so the Gauss coefficients are calculated from measurements of the northward (X), eastward (Y), and vertically downward (Z) components of the magnetic field at or above the Earth's surface (Fig. 7.1(a)). These components are related to other geomagnetic elements, such as the horizontal field (H), total field (T), angle of inclination (I), and angle of declination (D), as illustrated in Fig. 7.1(b). The field components in spherical polar coordinates are

$$X = -B_\theta = \frac{1}{r}\frac{\partial W}{\partial \theta}\bigg|_{r=R} \tag{7.17}$$

$$Y = B_\phi = \frac{-1}{r\sin\theta}\frac{\partial W}{\partial \phi}\bigg|_{r=R} \tag{7.18}$$

$$Z = -B_r = \frac{\partial W}{\partial r}\bigg|_{r=R} \tag{7.19}$$

The differentiations, after evaluating on the Earth's surface at $r = R$, result in the following set of equations involving the unknown Gauss coefficients:

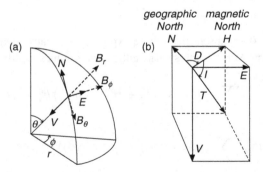

Fig. 7.1. (a) Relationship between the north (X), east (Y), and vertical (Z) components of the geomagnetic field and the spherical polar components B_r, B_θ, and B_ϕ. (b) The field may be described by the X, Y, and Z components, or by its intensity (T), declination (D), and inclination (I). A magnetic compass aligns with the horizontal component H, which is directed towards magnetic north.

$$X = \sum_{n=1}^{\infty} \sum_{m=0}^{n} \left(\{g_n^m + G_n^m\}\cos(m\phi) + \{h_n^m + H_n^m\}\sin(m\phi) \right) \frac{\partial}{\partial\theta} P_n^m(\cos\theta)$$

$$(7.20)$$

$$Y = \sum_{n=1}^{\infty} \sum_{m=0}^{n} \left(\{g_n^m + G_n^m\}\sin(m\phi) - \{h_n^m + H_n^m\}\cos(m\phi) \right) \frac{m}{\sin\theta} P_n^m(\cos\theta)$$

$$(7.21)$$

$$Z = -\sum_{n=1}^{\infty} \sum_{m=0}^{n} \left(\{(n+1)g_n^m - nG_n^m\}\cos(m\phi) \right.$$
$$\left. + \{(n+1)h_n^m - nH_n^m\}\sin(m\phi) P_n^m(\cos\theta) \right.$$

$$(7.22)$$

Note that the Gauss coefficients have the same dimensions as the magnetic field **B**, namely tesla. The tesla is a large magnetic field, so the geomagnetic field intensity and the Gauss coefficients are usually expressed in nanotesla (1 nT = 10^{-9} T). In the north and east components the Gauss coefficients occur as $(g_n^m + G_n^m)$ and $(h_n^m + H_n^m)$, and therefore the horizontal components alone do not allow separation of the external and internal parts. However, the Gauss coefficients occur in a different combination in the vertical field, and by virtue of this the external and internal fields can be separated.

In theory, the summations are over an infinite number of terms, but in practice they are truncated after a certain degree N. The coefficients h_n^0 and H_n^0 do not exist, because $\sin(m\phi) = 0$ for $m = 0$, and these terms make no contribution to the potential. For $n = 1$ there are three coefficients for the internal field (g_1^0, g_1^1, h_1^1) and three for the external field (G_1^0, G_1^1, H_1^1). Similarly, there are five of each for $n = 2$, and in general $2(2n + 1)$ for degree n. The total number of coefficients S_n up to and including order N for each part of the field is

$$S_N = [2(1) + 1] + [2(2) + 1] + [2(3) + 1] + \cdots + [2(N) + 1]$$
$$= 2(1 + 2 + 3 + \cdots + N) + N \qquad (7.23)$$

The sum of the first N natural numbers is $N(N + 1)/2$, so the number of coefficients up to degree and order N of the internal field is $N(N + 2)$. The same number is obtained for the external field. Thus separation requires knowing the field values at a minimum of $2N(N + 2)$ stations.

From 1835 to 1841 Carl Friedrich Gauss and Wilhelm Weber organized the semi-continuous (every 5 minutes, 24 hr/day) acquisition of data from up to 50 magnetic observatories distributed worldwide, albeit unevenly. Gauss in 1839 carried out the first analysis of the geomagnetic field up to degree and order 4, and established that it is dominantly of internal origin; the coefficients of the external field are small compared with those of the internal field, and may to a first approximation be neglected. The potential of the internal field is given by (7.14).

Magnetic field components have historically been measured and recorded at geomagnetic observatories. A drawback of the data from observatories is their uneven geographic distribution. A superior global coverage has been obtained during the last decades with the addition of data from satellites. The coefficients of the modern geomagnetic field have now been evaluated reliably up to degree and order 13. The data are updated and published regularly as the coefficients of the International Geomagnetic Reference Field (IGRF). The coefficients up to degree and order 3, corresponding to the dipole, quadrupole, and octupole components of the field at the Earth's surface are listed in Table 7.1 for some selected field models. The terms with $n = 1$ describe a dipole field; the higher-order terms with $n \geq 2$ are referred to collectively as the non-dipole field.

Table 7.1. *Dipole ($n = 1$), quadrupole ($n = 2$), and octupole ($n = 3$) Gauss–Schmidt coefficients from some historical field analyses. The coefficients DGRF are for Definitive Geomagnetic Reference Fields that will not be modified further. Details of the construction of the International Geomagnetic Reference Field IGRF 2010 are given in Finlay* et al. *(2010).*

	Epoch and source					
	1835, Gauss, in 1839	1885, Schmidt, in 1895	1922, Dyson and Furner (1923)	1965, DGRF	1985, DGRF	2010, IGRF
g_1^0	−32,350	−31,730	−30,920	−30,334	−29,873	−29,496.5
g_1^1	−3,110	−2,360	−2,260	−2,119	−1,905	−1,585.9
h_1^1	6,250	5,990	5,920	5,776	5,500	4,945.1
g_2^0	510	−520	−890	−1,662	−2,072	−2,396.6
g_2^1	2,920	2,830	2,990	2,997	3,044	3,026.0
h_2^1	120	−720	−1,240	−2,016	−2,197	−2,707.7
g_2^2	−20	680	1,440	1,594	1,687	1,668.6
h_2^2	1,570	1,500	840	114	−306	−575.4
g_3^0	–	940	1,140	1,297	1,296	1,339.7
g_3^1	–	−1,230	−1,650	−2,038	−2,208	−2,326.3
h_3^1	–	−300	−460	−404	−310	−160.5
g_3^2	–	1,430	1,200	1,292	1,247	1,231.7
h_3^2	–	30	120	240	284	251.7
g_3^3	–	400	880	856	829	634.2
h_3^3	–	680	230	−165	−297	−536.8

7.3 The Earth's dipole magnetic field

The dominant component of the Earth's surface magnetic field is the dipole component. The axis of the dipole is inclined to the rotation axis, thus it can be separated into an axial dipole and two orthogonal equatorial dipoles. As we will see, shifting these dipoles from the center of the Earth generates higher-order components in the geomagnetic potential.

7.3.1 The geocentric axial dipole

Each term in the geomagnetic potential (7.14) represents the potential of a particular pole configuration. The potential described by the largest coefficient, g_1^0, is

$$W_1^0 = \frac{R^3 g_1^0}{r^2} P_1^0(\cos\theta) = \frac{R^3 g_1^0 \cos\theta}{r^2} \qquad (7.24)$$

Comparison with (7.3) shows that this is the potential at distance r from the midpoint of a magnetic dipole and at angle θ from the dipole axis. In Earth coordinates this is the potential at co-latitude θ of a geocentric dipole aligned with the rotation axis and pointing to the north pole with magnetic moment m given by

$$m = \frac{4\pi R^3}{\mu_0} g_1^0 \qquad (7.25)$$

The magnetic field of an axial dipole is horizontal at the equator (see (7.4) and (7.5)). Its value at Earth's surface is

$$B_\theta = -\frac{1}{r}\frac{\partial}{\partial\theta}\left(\frac{R^3 g_1^0 \cos\theta}{r^2}\right)\bigg|_{r=R} = g_1^0 \sin\theta \qquad (7.26)$$

At the equator this is equal to g_1^0.

7.3.2 The geocentric inclined dipole

The coefficients of degree $n = 1$ and order $m = 1$ also have an inverse-square dependence on distance, so g_1^1 and h_1^1 too must represent dipoles. The combined potential of the dipole terms is

$$W_1 = R\left(\frac{R}{r}\right)^2 \left(g_1^0 P_1^0(\cos\theta) + (g_1^1 \cos\phi + h_1^1 \sin\phi)P_1^1(\cos\theta)\right) \qquad (7.27)$$

$$W_1 = R\left(\frac{R}{r}\right)^2 \left(g_1^0 \cos\theta + g_1^1 \cos\phi \sin\theta + h_1^1 \sin\phi \sin\theta\right) \qquad (7.28)$$

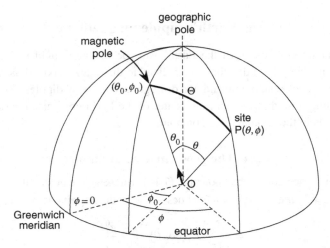

Fig. 7.2. Angular relationships pertaining to the computation of the potential of an inclined geocentric magnetic dipole.

Consider now the direction cosines of a line OP inclined at angle θ to the reference axis and at angle ϕ to the reference axis $\phi = 0$, as in Fig. 7.2. The direction cosines (α, β, γ) of OP are

$$\alpha = \sin \theta \cos \phi$$
$$\beta = \sin \theta \sin \phi \qquad (7.29)$$
$$\gamma = \cos \theta$$

Suppose the axis of a magnetic dipole to be inclined at angle θ_0 to the z-axis and at angle ϕ_0 to the reference axis $\phi = 0$. The direction cosines $(\alpha_0, \beta_0, \gamma_0)$ of the dipole axis are

$$\alpha_0 = \sin \theta_0 \cos \phi_0$$
$$\beta_0 = \sin \theta_0 \sin \phi_0 \qquad (7.30)$$
$$\gamma_0 = \cos \theta_0$$

If Θ is the angle between OP and the dipole axis, and r the distance of P from the dipole center, the magnetic potential at P is

$$W_1 = \frac{\mu_0 m}{4\pi r^2} \cos \Theta = \frac{\mu_0 m}{4\pi r^2} (\alpha\alpha_0 + \beta\beta_0 + \gamma\gamma_0) \qquad (7.31)$$

The components of the dipole moment m along the reference axes (Fig. 7.3) are

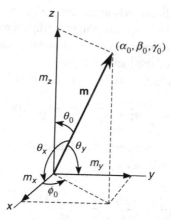

Fig. 7.3. Relationship between the Cartesian components and direction cosines of a magnetic dipole **m**, which is inclined at angle θ_0 to the rotation axis and has an azimuth ϕ_0 in the equatorial meridian.

$$m_x = m \cos \theta_x = m\alpha_0$$
$$m_y = m \cos \theta_y = m\beta_0 \qquad (7.32)$$
$$m_z = m \cos \theta_0 = m\gamma_0$$

The potential of the inclined dipole becomes

$$W_1 = \frac{\mu_0}{4\pi r^2} \left(\alpha m_x + \beta m_y + \gamma m_z \right) \qquad (7.33)$$

Using the relationships in (7.29), the potential of the inclined dipole is

$$W_1 = \frac{\mu_0}{4\pi r^2} \left(m_z \cos \theta + m_x \cos \phi \sin \theta + m_y \sin \phi \sin \theta \right) \qquad (7.34)$$

On equating individual terms with the expression for the potential using Gauss coefficients (7.28) it is evident that the coefficients g_1^1 and h_1^1 represent orthogonal dipoles in the equatorial plane. The equatorial dipole components are

$$m_x = \frac{4\pi R^3}{\mu_0} g_1^1 \qquad (7.35)$$

$$m_y = \frac{4\pi R^3}{\mu_0} h_1^1 \qquad (7.36)$$

The axial component of the dipole is

$$m_z = \frac{4\pi R^3}{\mu_0} g_1^0 \qquad (7.37)$$

The points where the dipole axis intersects the Earth's surface are called the geomagnetic poles (Fig. 7.2). At these points the dipole magnetic field is normal to the surface. The geomagnetic poles are antipodal to each other, because they lie at the opposite ends of the inclined axis. The co-latitude θ_0 of the pole is equal to the tilt of the inclined axis. From (7.30) and (7.32)

$$m \sin \theta_0 = \sqrt{m_x^2 + m_y^2} = \frac{4\pi}{\mu_0} R^3 \sqrt{\left(g_1^1\right)^2 + \left(h_1^1\right)^2} \qquad (7.38)$$

Together with the axial component, this defines the tilt θ_0 of the dipole axis, which is also the co-latitude of its pole:

$$\tan \theta_0 = \frac{m \sin \theta_0}{m \cos \theta_0} = \frac{\sqrt{\left(g_1^1\right)^2 + \left(h_1^1\right)^2}}{g_1^0} \qquad (7.39)$$

The components of the dipole moment in the equatorial plane, m_x and m_y, define the longitude ϕ_0 of the pole. From (7.35) and (7.36)

$$\tan \phi_0 = \frac{\beta_0}{\alpha_0} = \frac{m_y}{m_x} = \frac{h_1^1}{g_1^1} \qquad (7.40)$$

The dipole magnetic moment m is obtained by squaring and summing m_x, m_y, and m_z, giving

$$m = \frac{4\pi}{\mu_0} R^3 \sqrt{\left(g_1^0\right)^2 + \left(g_1^1\right)^2 + \left(h_1^1\right)^2} \qquad (7.41)$$

Analysis of the geomagnetic field for epoch 2010 (Finlay *et al.*, 2010) locates the north geomagnetic pole at 80.08 °N, 287.78 °E and the south geomagnetic pole at 80.08 °S, 107.78 °E. The places where the total magnetic field of the Earth is normal to the surface are the magnetic dip poles. The total field is expressed by all the terms in (7.14). Because of the non-dipole components the magnetic dip poles are not antipodal; also, because of secular variation (Section 7.4) the pole locations change slowly with time. For epoch 2010, the north dip pole was at 85.01 °N, 227.34 °E; the south dip pole was at 64.43 °S, 137.32 °E, which is outside the Antarctic Circle.

7.3.3 Axial dipole with axial offset

The terms with $n = 2$ are referred to as the quadrupole component of the field. However, one must keep in mind that the multipole expression of the magnetic field is a mathematical convenience that simply allows us to subdivide it for convenient reference. That is, just as there are no physical magnetic dipoles

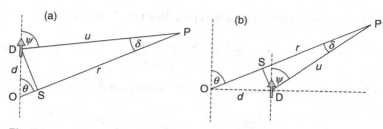

Fig. 7.4. (a) Geometry for calculation of the potential at P of an axial magnetic dipole at D, displaced a distance d along the rotation axis from the Earth's center at O. (b) The similar case of an axial magnetic dipole displaced in the equatorial plane.

inside the Earth, there are also no quadrupoles; a complex system of electric currents deep in the Earth causes the magnetic phenomena that we measure. The $n = 2$ coefficients are responsible for an offset of the magnetic dipole from the Earth's center. This can be shown as follows.

Let the axial magnetic dipole be displaced a small distance d along the dipole axis, as in Fig. 7.4(a). The position P is now a distance u from the center of the dipole at D, and the line DP makes an angle ψ with the dipole axis. The dipole potential at P is now

$$W = \frac{\mu_0}{4\pi} \frac{m \cos \psi}{u^2} \tag{7.42}$$

The line DP makes a small angle δ with the radius OP of length r. In the triangle ODP $\psi = \theta + \delta$, so

$$\cos \psi = \cos(\theta + \delta) = \cos \theta \cos \delta - \sin \theta \sin \delta \tag{7.43}$$

$$u^2 = r^2 + d^2 - 2rd\cos \theta \approx r^2 \left(1 - 2\frac{d}{r} \cos \theta \right) \tag{7.44}$$

In the triangle SDP, created by drawing DS perpendicular to OP,

$$\sin \delta = \frac{DS}{u} = \frac{d \sin \theta}{u} \tag{7.45}$$

For a very small displacement $d \ll r$, the distances r and u are almost equal, so the following relationships are approximately true to first order:

$$\sin \delta \approx \frac{d \sin \theta}{r} \quad \text{and} \quad \cos \delta \approx 1 \tag{7.46}$$

$$\cos \psi \approx \cos \theta - \frac{d}{r} \sin^2 \theta \tag{7.47}$$

The potential of the axially displaced dipole may now be written

$$W \approx \frac{\mu_0 m}{4\pi r^2} \frac{(\cos\theta - d/r\sin^2\theta)}{(1 - 2d/r\cos\theta)} \tag{7.48}$$

Using the binomial expansion and truncating it after the first order in d/r,

$$W \approx \frac{\mu_0 m}{4\pi r^2} \left(\cos\theta - \frac{d}{r}\sin^2\theta\right)\left(1 + 2\frac{d}{r}\cos\theta\right) \tag{7.49}$$

$$W \approx \frac{\mu_0 m}{4\pi r^2} \left(\cos\theta - \frac{d}{r}\sin^2\theta + 2\frac{d}{r}\cos^2\theta\right) \tag{7.50}$$

$$W \approx \frac{\mu_0 m}{4\pi r^2}\cos\theta + \frac{\mu_0 m d}{4\pi r^3}\left(3\cos^2\theta - 1\right) \tag{7.51}$$

$$W \approx \frac{\mu_0 m}{4\pi r^2} P_1^0(\cos\theta) + \frac{\mu_0(2md)}{4\pi r^3} P_2^0(\cos\theta) \tag{7.52}$$

The first term is the potential of a geocentric axial dipole; the second term is that of a geocentric axial quadrupole. An axial displacement of the dipole is equivalent to introducing the quadrupole term. The two terms are equivalent to the g_1^0 and g_2^0 terms in (7.14) for the multipole expansion of the potential.

7.3.4 Axial dipole with equatorial offset

To determine the effect of displacing the center of the axial dipole in the equatorial plane, we use the same approach as in the previous section. The geometry is as in Fig. 7.4(b) and the potential at P is, as before,

$$W = \frac{\mu_0}{4\pi}\frac{m\cos\psi}{u^2} \tag{7.53}$$

With reference to the triangle ODP, we now have $\psi = \theta - \delta$, so

$$\cos\psi = \cos(\theta - \delta) = \cos\theta\cos\delta + \sin\theta\sin\delta \tag{7.54}$$

$$u^2 = r^2 + d^2 - 2rd\cos\left(\frac{\pi}{2} - \theta\right) \approx r^2\left(1 - 2\frac{d}{r}\sin\theta\right) \tag{7.55}$$

For a very small displacement $d \ll r$, the distances r and u are almost equal, so to first order

$$\sin\delta \approx \frac{d\sin\theta}{r} \quad \text{and} \quad \cos\delta \approx 1 \tag{7.56}$$

$$\cos \psi \approx \cos \theta + \frac{d}{r} \sin^2 \theta \qquad (7.57)$$

In the triangle SDP, created by drawing DS perpendicular to OP,

$$\sin \delta = \frac{DS}{u} = \frac{d}{u} \sin\left(\frac{\pi}{2} - \theta\right) \approx \frac{d}{r} \cos \theta \qquad (7.58)$$

Using the binomial expansion and truncating it after the first order in d/r, the potential of the equatorially displaced dipole (7.53) may now be written

$$W \approx \frac{\mu_0 m}{4\pi r^2} \left(\cos \theta + \frac{d}{r} \sin^2 \theta\right) \bigg/ \left(1 - 2\frac{d}{r} \sin \theta\right)$$

$$\approx \frac{\mu_0 m}{4\pi r^2} \left(\cos \theta + \frac{d}{r} \sin^2 \theta\right)\left(1 + 2\frac{d}{r} \sin \theta\right) \qquad (7.59)$$

$$W \approx \frac{\mu_0 m}{4\pi r^2} \left(\cos \theta + 2\frac{d}{r} \sin \theta \cos \theta + \frac{d}{r} \sin^2 \theta\right) \qquad (7.60)$$

$$W \approx \frac{\mu_0 m}{4\pi r^2} \cos \theta + \frac{\mu_0 (2md)}{4\pi r^3} \sin \theta \cos \theta + \frac{\mu_0 (md)}{4\pi r^3} \sin^2 \theta \qquad (7.61)$$

Reference to Table 1.2 shows that the angular dependence of each term can be replaced by an associated Legendre polynomial, which gives

$$W \approx \frac{\mu_0 m}{4\pi r^2} P_1^0(\cos \theta) + \frac{\mu_0 (2md/3)}{4\pi r^3} P_2^1(\cos \theta) + \frac{\mu_0 (md/3)}{4\pi r^3} P_2^2(\cos \theta) \quad (7.62)$$

As before, the main term is the centered axial dipole. The additional terms result from the equatorial displacement, and are equivalent to the terms governed by coefficients g_2^1 and g_2^2 in (7.14).

7.3.5 Best-fitting eccentric inclined dipole

The best fit of a dipole to the observed magnetic field is obtained with an eccentric inclined dipole centered a few hundred kilometers from the center of the Earth (Box 7.1). To compute the offset of the dipole it is necessary to use all terms of degree and order $n \le 2$ in the multipole expansion of the potential. Using the Gauss coefficients for IGRF 2010 (Table 7.1), the location of the best-fitting eccentric inclined dipole has displacements $x_0 = -400$ km, $y_0 = 208$ km, $z_0 = 210$ km, $r_0 = 498$ km; i.e., it lies north of the equator under the North Pacific Ocean (Fig. 7.5) at 25 °N, 153 °E. The location of the eccentric dipole based on Quaternary and Recent paleomagnetic data and deep-sea cores (Creer et al.,

Box 7.1.   **The eccentric dipole**

The geomagnetic field is dominantly that of a dipole. The question
naturally arises as to the location of the dipole that best fits the present
field. Several methods of finding the optimum position have been
summarized by Lowes (1994). The most commonly used is a method that
was developed in 1934 by A. Schmidt, which yields the equations below
(Schmidt, 1934).

   The tilt of the dipole axis is determined by the Gauss coefficients of first
degree, $n = 1$. The best-fitting dipole is not centered at the center of the
Earth but is displaced to a position with coordinates (x_0, y_0, z_0), where z_0 is
the shift of the dipole center along the rotation axis, x_0 the shift in the
direction of the Greenwich meridian, and y_0 the shift orthogonal to these
displacements. The displacements can be determined approximately using
all Gauss coefficients with $n \le 2$; a more exact solution requires the $n = 3$
coefficients as well. The following equations describe the location of the
dipole center in a spherical Earth with radius R for $n \le 2$:

$$x_0 = R(L_1 - g_1^1 E)/(3m^2)$$
$$y_0 = R(L_2 - h_1^1 E)/(3m^2)$$
$$z_0 = R(L_0 - g_1^0 E)/(3m^2)$$
$$m^2 = \left(g_1^0\right)^2 + \left(g_1^1\right)^2 + \left(h_1^1\right)^2$$
$$E = \left(L_0 g_1^0 + L_1 g_1^1 + L_2 h_1^1\right)/(4m^2)$$
$$L_0 = 2g_1^0 g_2^0 + \left(g_1^1 g_2^1 + h_1^1 h_2^1\right)\sqrt{3}$$
$$L_1 = -g_1^1 g_2^0 + \left(g_1^0 g_2^1 + g_1^1 g_2^2 + h_1^1 h_2^2\right)\sqrt{3}$$
$$L_2 = -h_1^1 g_2^0 + \left(g_1^0 h_2^1 - h_1^1 g_2^2 + g_1^1 h_2^2\right)\sqrt{3}$$

The displacement r_0 of the center of the eccentric inclined dipole from the
center of the Earth is

$$r_0 = \sqrt{(x_0)^2 + (y_0)^2 + (z_0)^2}$$

1973) was found to be offset by about 200 km in the same direction, suggesting
the existence of persistent non-axial components in the global field.

   It is important to remember that the multipole method of expressing the
geomagnetic potential is a mathematical convenience. In reality there are no

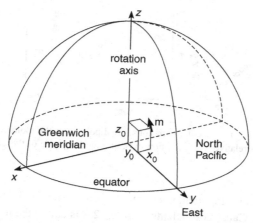

Fig. 7.5. The location of the best-fitting eccentric dipole for IGRF 2010 is offset into the northern hemisphere and the Pacific hemisphere. The orientation of the dipole is not changed by the offset.

dipoles, quadrupoles, or other multipoles. However, these concepts provide a convenient way of visualizing the geometry of parts of the field. As noted above, a displacement of a dipole from the center of coordinates creates higher-order terms in the multipole expansion. Thus it is possible to model the field with a moderate number of displaced dipoles. If each dipole corresponds to a current loop, this type of model may be physically more realistic. However, it is not practical for a mathematical description of the field.

7.4 Secular variation

The Gauss coefficients are not constants but change slowly with time, a phenomenon known as the secular variation of the field. Both the dipole and the non-dipole parts of the field exhibit secular variations. The dipole secular variations can be illustrated graphically by plotting the strength of the dipole magnetic moment and the orientation of the dipole axis, expressed as the latitude and longitude of a geomagnetic pole (Fig. 7.6). The timescale of dipole secular variations is of the order of thousands of years. The strength of the dipole magnetic moment has declined steadily over the past 150 years, during which observatory measurements of the field have been made. In the same time interval, the tilt of the dipole axis changed little until about 1960, but has since been decreasing. Similarly, the longitude of the geomagnetic pole was steady until the middle of the twentieth century, but has since been decreasing; this corresponds to a westward motion of the dipole axis around the rotation axis.

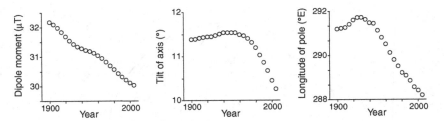

Fig. 7.6. Geomagnetic secular variations: the dipole magnetic moment, the tilt of the dipole axis relative to the rotation axis, and the longitude of the geomagnetic pole.

When the dipole component is subtracted from the total field, the remainder – described by the Gauss coefficients with $n \geq 2$ – is called the non-dipole field. Maps of the non-dipole field are characterized by large positive and negative anomalies that can have amplitudes amounting to a large fraction of the dipole field. These anomalies have a cell-like appearance, and change position and intensity with time. The non-dipole field has a standing (stationary) part, which exhibits intensity fluctuations without significant displacement, and a drifting (mobile) part. The best-known feature is a westward drift of many of the mapped cells at an average rate of about 0.3° per year.

7.5 Power spectrum of the internal field

The depth of the sources of the geomagnetic field of internal origin can be determined from the power spectrum of the Gauss coefficients. The power (or energy density) $\Re_n$ associated with the coefficients of degree n at the Earth's surface is given by Lowes (1966, 1974):

$$\Re_n = (n+1) \sum_{m=0}^{n} \left(\left(g_n^m\right)^2 + \left(h_n^m\right)^2 \right) \qquad (7.63)$$

The term of degree n in the geomagnetic potential varies with radial distance r as $r^{-(n+1)}$, so the strength of the field varies as $r^{-(n+2)}$. The power, or energy density, is proportional to the square of the amplitude, and thus varies as $r^{-2(n+2)}$. If the coefficients g_n^m and h_n^m have been determined on the surface of a sphere of radius r, the power spectrum on a surface of radius R closer to the center of the Earth is found by augmenting the spectrum by the ratio $(r/R)^{2(n+2)}$. The process is called downward continuation. The power spectrum on the surface of radius R is then given by

$$\Re_n(R) = \left(\frac{r}{R}\right)^{2(n+2)} \Re_n(r) \tag{7.64}$$

$$\Re_n(R) = (n+1)\left(\frac{r}{R}\right)^{2(n+2)} \sum_{m=0}^{n}\left((g_n^m)^2 + (h_n^m)^2\right) \tag{7.65}$$

The satellite MAGSAT measured the magnetic field at an average altitude of 420 km, equal to a radial distance of $r = 6{,}791$ km. The large quantity of data allowed harmonic analysis up to degree $n = 63$. The power spectrum based on the Gauss coefficients derived from the MAGSAT data is shown in Fig. 7.7 (lower curve). The $n = 1$ dipole term lies disproportionately above the other terms. On a semi-logarithmic plot the data form two almost linear segments, above and below $n = 14$. The part of the spectrum with $n \leq 14$ is attributed mainly to sources in the core; the part with higher values of n arises from sources mainly in the crust; the signal above $n \approx 50$ was considered to be noise, which averaged 0.091 nT^2 per degree. The two parts of the spectrum overlap around the break in slope.

The upper curve in Fig. 7.7 shows the data after downward continuation to the Earth's surface (radius $R = 6{,}371$ km). Note that the slope of the line for

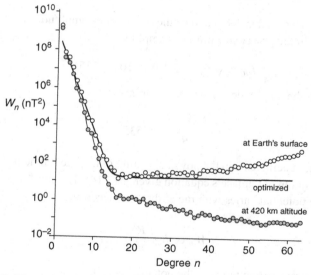

Fig. 7.7. The energy intensity associated with each degree of the spherical-harmonic analysis of the geomagnetic field, from measurements by the MAGSAT satellite at altitude 420 km, after reduction to the Earth's surface. Data source: Cain *et al.* (1989).

core sources ($n \leq 14$) is flatter than that at altitude 420 km. This suggests that if downward continuation is carried out to even deeper surfaces the slope might become zero. For $n > 15$ the slope of the line becomes positive. This is because downward continuation amplifies preferentially higher frequencies, including the noise inherent in the measured signal. When the noise is removed, the downward-continued spectrum at the Earth's surface is almost flat for $n > 15$ (the smooth curve in Fig. 7.7). The data after removing the average noise (and without the dipole term) can be fitted by a continuous curve with equation

$$\Re_n = 9.66 \times 10^8 (0.286)^n + 19.1 (0.996)^n \tag{7.66}$$

7.5.1 Estimation of the source depth of the main field

A method for estimating the approximate depth of the source layer of a magnetic or gravity anomaly is to assume that the power spectrum is "white" at that level (i.e., every part of the spectrum has the same amplitude). This can be applied to the non-dipole core field, for which

$$\Re_n = 9.66 \times 10^8 (0.286)^n \tag{7.67}$$

The power of a signal is defined to be the square of its amplitude. Thus the term of degree n in the power spectrum has amplitude

$$B_n = \sqrt{\Re_n} = 3.108 \times 10^4 (0.535)^n \tag{7.68}$$

The ratio of the amplitudes of successive terms is

$$\frac{B_{n+1}}{B_n} = 0.535 \tag{7.69}$$

The Gauss coefficients in the power spectrum of the internal field are defined from the solution of Laplace's equation given in (7.14). The amplitude of the nth term in the potential varies with radial distance according to

$$W_n \propto B_n \left(\frac{R}{r} \right)^{n+1} \tag{7.70}$$

The ratio of successive terms in the potential is then

$$\frac{W_{n+1}}{W_n} = \frac{B_{n+1}}{B_n} \left(\frac{R}{r} \right) = 0.535 \left(\frac{R}{r} \right) \tag{7.71}$$

If the power spectrum becomes white, then all terms in the potential are equal, $W_n = W_{n+1}$, and

$$r = 0.535R \tag{7.72}$$

This result locates the source layer of the non-dipole terms ($2 \leq n \leq 14$) at a radial distance of about 3,400 km. The radius of the core is 3,480 km, thus the source depth of the non-dipole terms is in the outer core, close to the core–mantle boundary.

The power spectrum at the Earth's surface, corrected for noise (solid line in Fig. 7.7), is almost flat above $n = 15$, signifying that the source layer of this part of the spectrum is very close to the surface and hence can be associated with crustal sources.

7.6 The origin of the internal field

William Gilbert's concept in 1600 of the Earth as a giant permanently magnetized sphere proved to be unrealistic in light of later knowledge of rock magnetic properties and the internal structure of the Earth. The magnetic field of a geocentric axial dipole is horizontal at the magnetic equator, where its strength B_e on the surface $r = R$ is

$$B_e = \frac{\mu_0}{4\pi} \frac{m \sin(\pi/2)}{R^3} = \frac{\mu_0}{4\pi R^3} m \tag{7.73}$$

The magnetization M is equal to the magnetic moment m per unit volume, so

$$B_e = \frac{\mu_0}{4\pi R^3} \frac{4\pi R^3}{3} M = \frac{\mu_0}{3} M \tag{7.74}$$

The equatorial field is equal to g_1^0 (i.e., ~30,000 nT), which gives a mean magnetization of 70 $A\,m^{-1}$. This greatly exceeds the magnetization of the most common strongly magnetized rocks (M is about 1 $A\,m^{-1}$ in basalt). Moreover, it does not take into account that the temperature inside the Earth soon exceeds the Curie temperature of magnetic minerals, above which no permanent magnetization is possible, so only the thin outer shell could be permanently magnetized. This would require an even greater magnetization than that calculated. Finally, the concept of a permanent magnet does not account for the observed secular variation of the magnetic field.

The experiments of Ampère and Ørsted in the early nineteenth century showed that magnetism was caused by electric currents. It is reasonable to assess whether the geomagnetic field has an electromagnetic origin.

7.6.1 Electromagnetic model

Maxwell's equations of electromagnetism (Appendix B) lead to an electromagnetic model for generation of the geomagnetic field in the fluid Earth's core. The electrical conductivity σ of the liquid-iron outer core is estimated to be about $5 \times 10^5 \ \Omega^{-1} \ m^{-1}$ (Stacey and Anderson, 2001), which makes it a good conductor. Any free charges would rapidly dissipate, so the free charge density ρ in Coulomb's law (Appendix B, part 1) is zero. A comparison of the magnitudes of the two terms on the right of Ampère's law (Appendix B, part 2) for a periodic variation with angular frequency $\omega = 2\pi/\tau$ gives

$$\frac{|\partial D/\partial t|}{|J|} = \frac{\varepsilon_0|\partial E/\partial t|}{\sigma|E|} = \frac{\varepsilon_0|i\omega E|}{\sigma|E|} = \frac{2\pi\varepsilon_0}{\tau\sigma} \tag{7.75}$$

The electric field constant is $\varepsilon_0 = 8.854 \times 10^{-12} \ C^2 \ N^{-1} \ m^{-2}$ and the approximate conductivity of the core is $\sigma = 5 \times 10^5 \ \Omega^{-1} \ m^{-1}$. For a period τ longer than a year (3.15×10^7 s), the ratio in (7.75) is less than 10^{-24}. Thus the displacement current $\partial D/\partial t$ can be ignored in the core. Maxwell's equations for the core become

$$\nabla \cdot \mathbf{E} = 0 \quad \text{(Coulomb's law)} \tag{7.76}$$

$$\nabla \times \mathbf{B} = \mu_0 \mathbf{J} \quad \text{(Ampère's law)} \tag{7.77}$$

$$\nabla \cdot \mathbf{B} = 0 \quad \text{(Gauss's law)} \tag{7.78}$$

$$\nabla \times \mathbf{E} = -\frac{\partial \mathbf{B}}{\partial t} \quad \text{(Faraday's law)} \tag{7.79}$$

Taking the curl of both sides of (7.77) gives

$$\nabla \times (\nabla \times \mathbf{B}) = \mu_0 \sigma (\nabla \times \mathbf{E}) \tag{7.80}$$

Substituting on the right from (7.79) gives

$$\nabla \times (\nabla \times \mathbf{B}) = -\mu_0 \sigma \frac{\partial \mathbf{B}}{\partial t} \tag{7.81}$$

Using the vector identity of (1.34), the left-hand side can be expanded, giving

$$\nabla(\nabla \cdot \mathbf{B}) - \nabla^2 \mathbf{B} = -\mu_0 \sigma \frac{\partial \mathbf{B}}{\partial t} \tag{7.82}$$

The first term can be eliminated because of Gauss's law, leaving

$$\nabla^2 \mathbf{B} = \mu_0 \sigma \frac{\partial \mathbf{B}}{\partial t} \tag{7.83}$$

$$\frac{\partial \mathbf{B}}{\partial t} = \frac{1}{\mu_0 \sigma} \nabla^2 \mathbf{B} = \eta_m \nabla^2 \mathbf{B} \tag{7.84}$$

This differential equation has the same form as the diffusion equation (6.66), and the parameter $\eta_m = 1/(\mu_0 \sigma)$ is called the *magnetic diffusivity*.

The magnetic field $\mathbf{B}$ must satisfy Gauss's law, having a solution such as

$$\mathbf{B} = -\nabla W + \nabla \times \mathbf{A} \tag{7.85}$$

In this solution the scalar potential W is the familiar solution of Laplace's equation, whereas $\mathbf{A}$ is a vector potential that must be added because of the vector identity that the divergence of the curl of a vector is always zero (see (1.33)). The scalar potential can be used for a magnetic field in a region that is free of electric currents (such as the description of the geomagnetic field using Gauss coefficients). A vector potential is appropriate to describe a field that arises from electric currents. If we insert this solution into (7.84) we get

$$\frac{\partial}{\partial t}(-\nabla W + \nabla \times \mathbf{A}) = \eta_m \nabla^2(-\nabla W + \nabla \times \mathbf{A}) \tag{7.86}$$

$$\nabla\left(\frac{\partial W}{\partial t} - \eta_m \nabla^2 W\right) = \frac{\partial}{\partial t}(\nabla \times \mathbf{A}) - \eta_m \nabla^2(\nabla \times \mathbf{A}) \tag{7.87}$$

Both sides of this equation have the same form as the thermal conductivity equation, if each side is set to zero. The solutions depend on space and time, and can be obtained by separating the variables with appropriate boundary conditions.

In a three-dimensional problem this can be complicated, but we can get an order-of-magnitude solution by considering a one-dimensional case. Let the scalar equation depend only on x and t,

$$\frac{\partial W}{\partial t} = \eta_m \frac{\partial^2 W}{\partial x^2} \tag{7.88}$$

This is a magnetic equivalent of the heat-conduction equation (Section 6.6.2). A possible solution is

$$W = W_0 \sin\left(2\pi n \frac{x}{L}\right) \cdot \exp(-t/\tau) \tag{7.89}$$

The quantity L is a length that is characteristic of the problem. It may be comparable to the size of the outer core, for example. The magnetic potential W decays exponentially; the quantity τ is a relaxation time, over which the field sinks to $1/e$ of its initial value. Upon inserting the solution into (7.88) and taking the fundamental mode of the distance dependence ($n = 1$) we get

$$-\frac{1}{\tau} W = -\frac{4\pi^2 \eta_{\mathrm{m}}}{L^2} W \tag{7.90}$$

This gives the relaxation time in terms of other core parameters:

$$\tau = \frac{\mu_0 \sigma L^2}{4\pi^2} \tag{7.91}$$

The electrical conductivity of the core is approximately $5 \times 10^5 \ \Omega^{-1} \ \mathrm{m}^{-1}$ and $\mu_0 = 4\pi \times 10^{-7} \ \mathrm{N} \ \mathrm{A}^{-2}$, so, taking a characteristic length $L = 2,000 \ \mathrm{km}$, the relaxation time τ is $6.4 \times 10^{10} \ \mathrm{s}$ or about 2,000 yr. In a time equal to 5τ an exponential function sinks to less than 1% of its initial value, so the magnetic field generated by a purely electromagnetic model would disappear in about 10,000 years. Magnetizations in ancient rocks show that the Earth has had a magnetic field since the Pre-Cambrian, i.e., for times on the order of 10^9 yr, so the electromagnetic model is inadequate. A satisfactory model must be capable of sustaining a magnetic field for this long.

A further mechanism is needed to regenerate the magnetic field and prevent it from diffusing away. This is provided by physical motion of the electrically conducting core fluid, which interacts with the magnetic field lines in the core. The mechanism is analogous to that of a dynamo, in which a coil of wire is moved through the field of a magnet to create an electric current in the wire. The process of generating the geomagnetic field by induction from the motion of the conducting core fluid is known as the *dynamo model*.

7.6.2 The magnetohydrodynamic model

When an electrical charge q moves with velocity $\mathbf{v}$ through a magnetic field $\mathbf{B}$, it experiences the Lorentz force $\mathbf{F}$, which is normal to the field and to the direction of motion (Appendix A3):

$$\mathbf{F} = q(\mathbf{v} \times \mathbf{B}) \tag{7.92}$$

In the case of the Earth's core it gives rise to an additional electric field $\mathbf{E_L}$ given by

$$\mathbf{E_L} = \frac{\mathbf{F}}{q} = \mathbf{v} \times \mathbf{B} \tag{7.93}$$

The total electric field experienced by the material of the core is now $\mathbf{E_t} = \mathbf{E} + \mathbf{E_L}$, and for Ohm's law we get

$$\mathbf{J} = \sigma \mathbf{E_t} = \sigma(\mathbf{E} + \mathbf{E_L}) = \sigma(\mathbf{E} + \mathbf{v} \times \mathbf{B}) \tag{7.94}$$

Ampère's equation (7.77) becomes

$$\nabla \times \mathbf{B} = \mu_0 \mathbf{J} = \mu_0 \sigma (\mathbf{E} + \mathbf{v} \times \mathbf{B}) \tag{7.95}$$

With the additional term we now proceed as for the electromagnetic model, taking the curl of both sides of the equation:

$$\nabla \times (\nabla \times \mathbf{B}) = \mu_0 \sigma ((\nabla \times \mathbf{E}) + (\nabla \times \mathbf{v} \times \mathbf{B})) \tag{7.96}$$

$$\nabla(\nabla \cdot \mathbf{B}) - \nabla^2 \mathbf{B} = \mu_0 \sigma \left(-\frac{\partial \mathbf{B}}{\partial t} + \nabla \times \mathbf{v} \times \mathbf{B} \right) \tag{7.97}$$

The first term is zero because of Gauss's law; rearranging the other terms gives

$$\frac{\partial \mathbf{B}}{\partial t} = \eta_{\mathrm{m}} \nabla^2 \mathbf{B} + (\nabla \times \mathbf{v} \times \mathbf{B}) \tag{7.98}$$

This is known as the *magnetohydrodynamic induction equation*. The constant η_{m} is the magnetic diffusivity, as before. As a result of the additional term on the right, the magnetic field no longer decays exponentially with time. The first term describes the tendency of the field to decay by diffusion; the second term provides additional energy to regenerate the field from the interaction of the field with the motion of the conducting fluid. The ratio of the terms on the right is called the magnetic Reynolds number, R_{m}, defined as

$$R_{\mathrm{m}} = \frac{|\nabla \times \mathbf{v} \times \mathbf{B}|}{|\eta_{\mathrm{m}} \nabla^2 \mathbf{B}|} \tag{7.99}$$

The magnetic Reynolds number is defined by analogy with fluid mechanics, where the Reynolds number is a property of a fluid that determines the predominance of laminar flow or turbulent flow. At low Reynolds numbers viscous forces are dominant, and the flow is laminar; at high Reynolds numbers inertial forces result in turbulent flow, which is less stable and typified by random eddies. For a magnetic Reynolds number $R_{\mathrm{m}} \ll 1$, the magnetic field simply diffuses away by ohmic dissipation as in the electromagnetic example discussed in the previous section. If $R_{\mathrm{m}} \gg 1$, the magnetic-field lines are carried along by the conducting fluid and the fluid motion predominates in the generation of the field.

We can use dimensional analysis to estimate the magnitude of R_{m} in the core. The dimension of a gradient is $[L]^{-1}$, we can write $[B]$ for the dimension of the field, and the magnetic diffusivity $\eta_{\mathrm{m}} = 1/(\mu_0 \sigma)$. Thus

$$R_{\mathrm{m}} = \frac{|\nabla \times \mathbf{v} \times \mathbf{B}|}{\eta_{\mathrm{m}} |\nabla^2 \mathbf{B}|} \approx \frac{\mu_0 \sigma [L]^{-1} [v][B]}{[L]^{-2}[B]} \tag{7.100}$$

$$R_{\mathrm{m}} = \mu_0 \sigma v L \tag{7.101}$$

The quantities v and L are not known precisely. L is an unspecified length assumed to be typical for a core motion; we may use the same value as before for the core, i.e., $L = 2{,}000$ km. The velocity v of the conducting fluid has been estimated from the westward motion of field features to be on the order of 10–20 km yr^{-1}, i.e., $v \approx 0.3$–0.6 mm s^{-1}. This gives a magnetic Reynolds number of about 250–500. Even slower motions of the core give $R_m \gg 1$, so to a first approximation we can ignore the diffusive term and write

$$\frac{\partial \mathbf{B}}{\partial t} = \nabla \times \mathbf{v} \times \mathbf{B} \qquad (7.102)$$

This equation would be exactly true for a material with infinite conductivity, but the finite conductivity of the core means that there is some leakage of the magnetic flux. However, the assumption of infinite conductivity allows deeper insight into the generation of the geomagnetic field.

7.6.3 The frozen-flux theorem

Let S be a surface bounded by a closed loop L in an electrically conducting fluid at time t, and let $\mathbf{B}(t)$ be a magnetic field cutting S (Fig. 7.8). If $d\mathbf{S}$ is an element of the surface area, the magnetic flux Φ_0 through S is

$$\Phi_0 = \int_S \mathbf{B}(t) \cdot d\mathbf{S} \qquad (7.103)$$

Suppose that the conducting fluid moves with velocity $\mathbf{v}$. In a short time increment Δt the loop is displaced through a small distance $d\mathbf{x} = \mathbf{v}\,\Delta t$. This defines a cylinder of volume V with a total surface area A, made up of (1) the bottom surface with

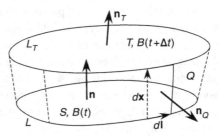

Fig. 7.8. Configuration for derivation of the "frozen-flux theorem." At time t the magnetic field $B(t)$ intersects a surface S moving with velocity $\mathbf{v}$ through a conducting fluid; at time $t + \Delta t$ the field has changed to $B(t + \Delta t)$ and the surface area has changed to T. Relative to the enclosed volume, the normal directions $\mathbf{n}_T$ and $\mathbf{n}_Q$ to surfaces T and Q are outward; the normal direction $\mathbf{n}$ to the bottom surface S is inward.

area S bounded by loop L, (2) the top surface with area T bounded by loop L_T, and (3) the side surfaces with area Q. During the elapsed time Δt the magnetic field itself changes to $\mathbf{B}(t + \Delta t)$. The flux Φ_2 through the top surface T is

$$\Phi_2 = \int_T \mathbf{B}(t + \Delta t) \cdot d\mathbf{S} \tag{7.104}$$

We can apply the divergence theorem (Section 1.6) and Gauss's law for magnetism to the volume V cut by the field lines of $\mathbf{B}$. At any time

$$\int_A \mathbf{B} \cdot d\mathbf{S} = \int_V (\nabla \cdot \mathbf{B}) dV = 0 \tag{7.105}$$

The integration on the left is the flux of the magnetic field through all the surfaces bounding the volume V. It can be written as the sum of the flux through each end surface plus the flux through the side surface: thus, at time $t + \Delta t$,

$$-\int_S \mathbf{B}(t + \Delta t) \cdot d\mathbf{S} + \int_T \mathbf{B}(t + \Delta t) \cdot d\mathbf{S} + \int_Q \mathbf{B}(t + \Delta t) \cdot d\mathbf{S} = 0 \tag{7.106}$$

The negative sign in the first term is necessary because the normal direction to each surface is outward, but we have defined the flux of the field to be inward across S and outward across T. On rearranging terms, the flux across the top surface T is given by

$$\Phi_2 = \int_T \mathbf{B}(t + \Delta t) \cdot d\mathbf{S} = \int_S \mathbf{B}(t + \Delta t) \cdot d\mathbf{S} - \int_Q \mathbf{B}(t + \Delta t) \cdot d\mathbf{S} \tag{7.107}$$

The change in flux has two causes: the first is the change in the magnetic field with time, and the second is the change of surface area through which the field passes. If the time Δt is short, we can write the first term on the right to first order as

$$\mathbf{B}(t + \Delta t) = \mathbf{B}(t) + \frac{\partial \mathbf{B}(t)}{\partial t} \Delta t \tag{7.108}$$

Upon inserting this into (7.107) we have

$$\Phi_2 = \int_S \mathbf{B}(t) \cdot d\mathbf{S} + \Delta t \int_S \frac{\partial \mathbf{B}(t)}{\partial t} \cdot d\mathbf{S} - \int_Q \mathbf{B}(t + \Delta t) \cdot d\mathbf{S} \tag{7.109}$$

The change in flux through the moving loop is

$$\Delta\Phi = \Phi_2 - \Phi_0 = \Delta t \int_S \frac{\partial \mathbf{B}(t)}{\partial t} \cdot d\mathbf{S} - \int_Q \mathbf{B}(t + \Delta t) \cdot d\mathbf{S} \qquad (7.110)$$

The flux through the side surfaces must now be calculated. In time Δt the displacement parallel to the local velocity vector of the fluid is $d\mathbf{x} = \mathbf{v}\,\Delta t$. Together with an incremental distance $d\mathbf{l}$ along the loop L, this displacement defines an element of the surface Q with area

$$d\mathbf{S} = d\mathbf{l} \times d\mathbf{x} = (d\mathbf{l} \times \mathbf{v})\Delta t \qquad (7.111)$$

Thus the magnetic flux across the side surface Q is

$$\int_Q \mathbf{B}(t + \Delta t) \cdot d\mathbf{S} = \Delta t \int_Q \mathbf{B}(t + \Delta t) \cdot (d\mathbf{l} \times \mathbf{v}) \qquad (7.112)$$

We can change the variable of integration by using the vector identity in (1.18). The surface integration over Q is converted into a linear integration along $d\mathbf{l}$, i.e., around the closed loop L:

$$\int_Q \mathbf{B}(t + \Delta t) \cdot d\mathbf{S} = \Delta t \int_L (\mathbf{v} \times \mathbf{B}(t + \Delta t)) \cdot d\mathbf{l} \qquad (7.113)$$

Now we again use (7.108) to replace $\mathbf{B}(t + \Delta t)$ by $\mathbf{B}(t)$ and its time-derivative:

$$\int_Q \mathbf{B}(t + \Delta t) \cdot d\mathbf{S} = \Delta t \int_L \left(\mathbf{v} \times \left(\mathbf{B}(t) + \frac{\partial \mathbf{B}(t)}{\partial t}\Delta t \right) \right) \cdot d\mathbf{l}$$

$$= \Delta t \int_L \left(\mathbf{v} \times \mathbf{B}(t) \right) \cdot d\mathbf{l} + (\Delta t)^2 \int_L \left(\mathbf{v} \times \frac{\partial \mathbf{B}(t)}{\partial t} \right) \cdot d\mathbf{l}$$

$$(7.114)$$

By inserting this expression into (7.110) we obtain the change in flux in time Δt:

$$\Delta\Phi = \Delta t \int_S \frac{\partial \mathbf{B}(t)}{\partial t} \cdot d\mathbf{S} - \Delta t \int_L \left(\mathbf{v} \times \mathbf{B}(t) \right) \cdot d\mathbf{l} - (\Delta t)^2 \int_L \left(\mathbf{v} \times \frac{\partial \mathbf{B}(t)}{\partial t} \right) \cdot d\mathbf{l}$$

$$(7.115)$$

On dividing throughout by Δt, we have

$$\frac{\Delta\Phi}{\Delta t} = \int_S \frac{\partial \mathbf{B}(t)}{\partial t} \cdot d\mathbf{S} - \int_L (\mathbf{v} \times \mathbf{B}(t)) \cdot d\mathbf{l} - \Delta t \int_L \left(\mathbf{v} \times \frac{\partial \mathbf{B}(t)}{\partial t} \right) \cdot d\mathbf{l} \quad (7.116)$$

The rate of change of magnetic flux is the limit of this expression as Δt tends to zero; the final term disappears and

$$\frac{d\Phi}{dt} = \lim_{\Delta t=0}\left(\frac{\Delta\Phi}{\Delta t}\right) = \int_S \frac{\partial \mathbf{B}(t)}{\partial t} \cdot d\mathbf{S} - \int_L (\mathbf{v} \times \mathbf{B}(t)) \cdot d\mathbf{l} \qquad (7.117)$$

The integral around the closed loop L can be converted into an integral over the open bounded surface S by applying Stokes' theorem (Section 1.7):

$$\int_L (\mathbf{v} \times \mathbf{B}(t)) \cdot d\mathbf{l} = \int_S (\nabla \times \mathbf{v} \times \mathbf{B}(t)) \cdot d\mathbf{S} \qquad (7.118)$$

The rate of change of magnetic flux through the closed loop L is therefore

$$\frac{d\Phi}{dt} = \int_S \left[\frac{\partial \mathbf{B}(t)}{\partial t} - \left(\nabla \times \mathbf{v} \times \mathbf{B}(t)\right)\right] \cdot d\mathbf{S} \qquad (7.119)$$

If the electrical conductivity of the moving fluid is infinite, the approximation in (7.102) applies, and the expression in brackets is zero. Therefore,

$$\frac{d\Phi}{dt} = 0 \qquad (7.120)$$

and

$$\Phi = \int_S \mathbf{B}(t) \cdot d\mathbf{S} = \text{constant} \qquad (7.121)$$

This result states that the magnetic flux in a fluid with infinite electrical conductivity does not change as the fluid moves. This is known as the frozen-flux (or frozen-in-flux) theorem. It was formulated in 1943 by H. Alfvén, a Swedish physicist, for an electrically conductive plasma (such as the solar wind). The theorem can be applied as an approximation for any conducting fluid with a high magnetic Reynolds number, such as the Earth's liquid core. It describes how, in an ideal case, magnetic field lines are trapped by the high conductivity and compelled to move with the fluid. As a result, fluid motions in the core, in particular thermally and compositionally driven convection, provide the energy source and feedback mechanism for a self-sustaining magnetic field.

FURTHER READING

Campbell, W. H. (2003). *Introduction to Geomagnetic Fields*. Cambridge: Cambridge University Press, 337 pp.

Gubbins, D. and Herrero-Bervera, E. (2007). *Encyclopedia of Geomagnetism and Paleomagnetism*. Dordrecht: Springer, 1,054 pp.

Merrill, R. T., McElhinny, M. W., and McFadden, P. L. (1996). *The Magnetic Field of the Earth: Paleomagnetism, the Core, and the Deep Mantle*. San Diego, CA: Academic Press, 527 pp.

8

Foundations of seismology

Our knowledge of Earth's internal structure has been obtained from detailed analysis of the travel-times of seismic waves in the Earth. A standard model of the layered interior – PREM, the Preliminary Reference Earth Model (Dziewonski and Anderson, 1981) – that gives the variations with depth of seismic velocities, density, pressure, and elastic parameters has been derived. This chapter handles the dependence of seismic-wave velocities on the elastic properties of the medium in which they are transmitted.

The propagation of a seismic wave takes place by infinitesimal elastic displacements of the material it passes through. An elastic displacement is reversible, i.e., after the disturbing force has been removed the material returns to its original condition. The elastic properties and density of the material determine the type of wave that passes through it, and the speed with which the wave travels.

8.1 Elastic deformation

Elastic deformation is governed by *Hooke's law*, which was formulated in the seventeenth century on the basis of empirical observations. These are illustrated by the deformation of a rod of length x and cross-sectional area A, which extends by an amount δx due to an applied force F (Fig. 8.1). In an elastic deformation the fractional increase in length ($\delta x/x$) is directly proportional to the applied force F and inversely proportional to its cross-section A:

$$\frac{\delta x}{x} \propto \frac{F}{A} \tag{8.1}$$

Stress and strain are defined for a small volume of a continuous medium as limiting cases when the volume shrinks to zero, i.e., when both the length x and the cross-sectional area A become very small. The limit of the force per unit area (F/A) is the *stress*, σ, which has the units of pressure (pascal):

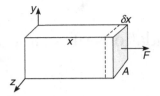

Fig. 8.1. Extension of a rod of length x and cross-sectional area A due to an applied force F.

$$\sigma = \lim_{A \to 0} \left(\frac{F}{A} \right) \tag{8.2}$$

The limit of the fractional change in dimension ($\delta x/x$) is the *strain*, ε, which is dimensionless:

$$\varepsilon = \lim_{x \to 0} \left(\frac{\delta x}{x} \right) \tag{8.3}$$

Hooke's law states that in an elastic deformation the stress and strain are proportional to each other:

$$\sigma \propto \varepsilon \tag{8.4}$$

The law describes the initial deformation of a material; the stress–strain relationship is linear, and the behavior is said to be *perfectly elastic.* If the stress increases continuously, the linearity breaks down, but the behavior is still elastic and no permanent deformation results (Fig. 8.2). Eventually the limit of elastic behavior is reached, permanent deformation results, and finally failure occurs. The propagation of seismic waves takes place within the elastic range of behavior.

8.2 Stress

The forces acting on an elastic body can be divided into body forces (e.g., gravity, centrifugal force) and surface forces (e.g., pressure, tension, and shear). Imagine a small volume δV bounded by a surface S within a continuous larger body of uniform density ρ. The body forces acting on δV (including inertial forces) produce acceleration of δV and of the body as a whole. The material surrounding δV exerts inward forces on the surface S; to maintain equilibrium, equal and opposite surface forces act outwards across S. They cause the small volume to change shape and define the state of stress in the body.

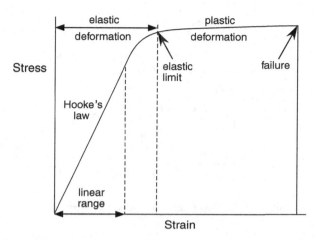

Fig. 8.2. Hypothetical stress–strain relationship, showing the regions of elastic and plastic deformation, and the linear range within which Hooke's law holds.

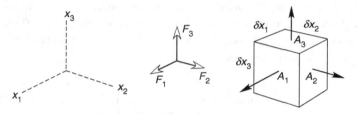

Fig. 8.3. Definitions of the quantities involved in calculating the components of stress caused by force components F_1, F_2, and F_3 acting on the sides of a small rectangular box with surface areas A_1, A_2, and A_3, respectively.

The definition of components of stress is illustrated for a small rectangular box. Let $\mathbf{F}$ be a force with components F_1, F_2, and F_3 referred to orthogonal Cartesian coordinate axes x_1, x_2, and x_3, respectively. $\mathbf{F}$ acts upon the surfaces of a small rectangular box with sides parallel to the reference axes (Fig. 8.3). The direction of each component of $\mathbf{F}$ is normal to one of the surfaces and tangential to the other two. The orientation of each surface is specified by its outward normal, and the respective areas are A_1, A_2, and A_3.

The component of force F_1 normal to the surface A_1 produces a *normal stress*, denoted σ_{11}. The components F_2 and F_3 tangential to the surface A_1 result in *shear stresses* σ_{12} and σ_{13}. The three components of stress acting on the surface A_1 are defined as

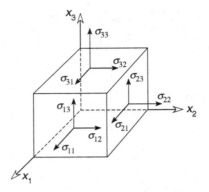

Fig. 8.4. Definition of the components of normal and shear stress.

$$\sigma_{11} = \lim_{A_1 \to 0} \left(\frac{F_1}{A_1}\right), \qquad \sigma_{12} = \lim_{A_1 \to 0} \left(\frac{F_2}{A_1}\right), \qquad \sigma_{13} = \lim_{A_1 \to 0} \left(\frac{F_3}{A_1}\right) \quad (8.5)$$

Similarly, the components of $\mathbf{F}$ acting on the surface A_2 define a normal stress σ_{22} and shear stresses σ_{21} and σ_{23}, while the components of $\mathbf{F}$ acting on the surface A_3 define a normal stress σ_{33} and shear stresses σ_{31} and σ_{32} (Fig. 8.4). The nine components σ_{kn} ($k = 1, 2, 3$; $n = 1, 2, 3$) form the elements of the *stress tensor*, which in matrix form is

$$\sigma_{kn} = \begin{pmatrix} \sigma_{11} & \sigma_{12} & \sigma_{13} \\ \sigma_{21} & \sigma_{22} & \sigma_{23} \\ \sigma_{31} & \sigma_{32} & \sigma_{33} \end{pmatrix} \quad (8.6)$$

In each case the first index of a stress element identifies the orientation of a surface and the second index identifies the component of force acting on the surface.

8.2.1 Symmetry of the stress tensor

Let the sides of the small rectangular box have lengths δx_1, δx_2, and δx_3 parallel to the reference axes (Fig. 8.5). For the box to be in static equilibrium, the sum of the forces on the box (which would displace it) must be zero, and the sum of the moments of the forces (which would rotate it) must also be zero. Consider first the balance of the moments acting on pairs of faces. The couple exerted about a line through the center of the box parallel to the x_3-axis by the shear stresses on the faces normal to x_1 (Fig. 8.5(a)) is (to first order, neglecting the second-order term in δx_1^2)

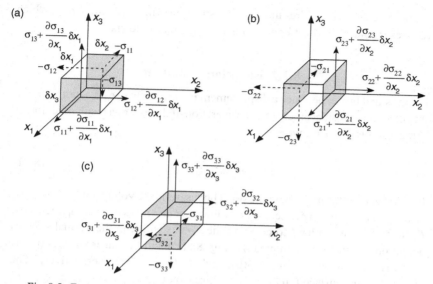

Fig. 8.5. Forces acting on the surfaces of a small rectangular box in the directions of (a) the x_1-axis, (b) the x_2-axis, and (c) the x_3-axis.

$$\left(\sigma_{12} + \frac{\partial \sigma_{12}}{\partial x_1} \delta x_1\right) A_1 \frac{\delta x_1}{2} + \sigma_{12} A_1 \frac{\delta x_1}{2} = \sigma_{12}\, \delta x_1\, A_1 = \sigma_{12}\, \delta x_1\, \delta x_2\, \delta x_3$$

$$= \sigma_{12}\, \delta V$$

$$(8.7)$$

A further couple is exerted about the x_3-axis by the shear stresses on the faces normal to x_2 (Fig. 8.5(b)). This acts in the opposite sense to the first couple and (also to first order) is equal to

$$\left(\sigma_{21} + \frac{\partial \sigma_{21}}{\partial x_2} \delta x_2\right) A_2 \frac{\delta x_2}{2} + \sigma_{21} A_2 \frac{\delta x_2}{2} = \sigma_{21}\, \delta x_2\, A_2 = \sigma_{21}\, \delta V \qquad (8.8)$$

The resulting couple about the x_3-axis is the difference between (8.7) and (8.8). For the box to be in equilibrium, the sum of the moments about the x_3-axis must be zero; therefore

$$(\sigma_{12} - \sigma_{21})\delta V = 0 \qquad (8.9)$$

This must be valid for any small volume δV; therefore,

$$\sigma_{12} = \sigma_{21} \qquad (8.10)$$

Similar evaluations of the moments about the x_1- and x_2-axes show, respectively, that $\sigma_{23} = \sigma_{32}$ and $\sigma_{31} = \sigma_{13}$. The equilibrium of moments acting on the elementary volume requires the stress tensor to be symmetric ($\sigma_{kn} = \sigma_{nk}$), which reduces the number of different elements in the matrix to six.

8.2.2 Equation of motion

Let the small box experience a displacement $\mathbf{u} = u_n \mathbf{e}_n$, where $\mathbf{e}_n$ is a unit vector in the direction of displacement. The acceleration of the box as a result of all forces acting on it is $\mathbf{a} = a_n \mathbf{e}_n$, where

$$a_n = \frac{\partial^2 u_n}{\partial t^2} \tag{8.11}$$

If the density of the material in the small box is ρ and the volume of the box is δV, its mass m is equal to $\rho\,\delta V$. Let the body force per unit mass have components F_1, F_2, and F_3. The resultant force along the x_1-axis is due to the normal stresses acting on the surfaces with area A_1 (Fig. 8.5(a)) and the shear stresses on the surfaces with areas A_2 (Fig. 8.5(b)) and A_3 (Fig. 8.5(c)), respectively. The resultant of the surface forces in the x_1-direction is

$$\left(\sigma_{11} + \frac{\partial \sigma_{11}}{\partial x_1}\delta x_1 - \sigma_{11}\right)A_1 + \left(\sigma_{21} + \frac{\partial \sigma_{21}}{\partial x_2}\delta x_2 - \sigma_{21}\right)A_2$$

$$+ \left(\sigma_{31} + \frac{\partial \sigma_{31}}{\partial x_3}\delta x_3 - \sigma_{31}\right)A_3$$

$$= \frac{\partial \sigma_{11}}{\partial x_1}\delta x_1(\delta x_2\,\delta x_3) + \frac{\partial \sigma_{21}}{\partial x_2}\delta x_2(\delta x_3\,\delta x_1) + \frac{\partial \sigma_{31}}{\partial x_3}\delta x_3(\delta x_1\,\delta x_2)$$

$$= \left(\frac{\partial \sigma_{11}}{\partial x_1} + \frac{\partial \sigma_{21}}{\partial x_2} + \frac{\partial \sigma_{31}}{\partial x_3}\right)\delta V \tag{8.12}$$

The equation of motion in the x_1-direction as a result of the inertial, body, and surface forces is

$$ma_1 = mF_1 + \left(\frac{\partial \sigma_{11}}{\partial x_1} + \frac{\partial \sigma_{21}}{\partial x_2} + \frac{\partial \sigma_{31}}{\partial x_3}\right)\delta V \tag{8.13}$$

$$\rho a_1 = \rho F_1 + \left(\frac{\partial \sigma_{11}}{\partial x_1} + \frac{\partial \sigma_{21}}{\partial x_2} + \frac{\partial \sigma_{31}}{\partial x_3}\right) \tag{8.14}$$

Similar expressions are obtained for the net forces along the x_2- and x_3-axes. Using the summation convention (where the repeated index implies the sum for $k = 1$, 2, and 3), we get the tensor equation

$$\rho a_n = \rho F_n + \frac{\partial \sigma_{kn}}{\partial x_k} \tag{8.15}$$

If the body force per unit mass F_n can be neglected, we can write the acceleration as in (8.11), and this equation reduces to the *homogeneous equation of motion*:

$$\rho \frac{\partial^2 u_n}{\partial t^2} = \frac{\partial \sigma_{kn}}{\partial x_k} \tag{8.16}$$

8.3 Strain

Let the vector $\mathbf{x}$ define a point P in an arbitrary body and let Q be another point of the body at an infinitesimal distance $\mathbf{y}$ from P, as in Fig. 8.6. In a general displacement of the body the point P is displaced to a new position P_1 by the vector $\mathbf{u}$, and Q is displaced to Q_1 by the vector $\mathbf{v}$. If the difference between the displacements is $d\mathbf{u}$, then

$$\mathbf{v} = \mathbf{u} + d\mathbf{u} = \mathbf{u} + \frac{\partial \mathbf{u}}{\partial x_1} y_1 + \frac{\partial \mathbf{u}}{\partial x_2} y_2 + \frac{\partial \mathbf{u}}{\partial x_3} y_3 \tag{8.17}$$

Here y_1, y_2, and y_3 are the components of $\mathbf{y}$ in the directions of the coordinates x_1, x_2, and x_3, respectively. In tensor notation

$$v_k = u_k + du_k = u_k + \frac{\partial u_k}{\partial x_n} y_n \tag{8.18}$$

The relationship is not changed if we subtract the term $\frac{1}{2} \, \partial u_n / \partial x_k$, and then add it back again, giving

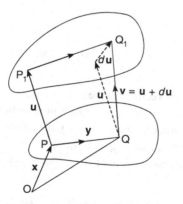

Fig. 8.6. Illustration of a general displacement of points in a medium. The point P is displaced to a new position P_1 by the vector $\mathbf{u}$ and Q is displaced to Q_1 by the vector $\mathbf{v}$.

$$v_k = u_k + \frac{1}{2}\left(\frac{\partial u_k}{\partial x_n} - \frac{\partial u_n}{\partial x_k}\right)y_n + \frac{1}{2}\left(\frac{\partial u_k}{\partial x_n} + \frac{\partial u_n}{\partial x_k}\right)y_n \qquad (8.19)$$

$$v_k = u_k + \varphi_{kn}y_n + \varepsilon_{kn}y_n \qquad (8.20)$$

The first term on the right-hand side of this equation represents a rigid-body *translation* of the entire body by the vector **u**. This takes place without internal deformation of the body.

The second term on the right contains the tensor φ_{kn}, whose elements are

$$\varphi_{kn} = \frac{1}{2}\left(\frac{\partial u_k}{\partial x_n} - \frac{\partial u_n}{\partial x_k}\right) \qquad (8.21)$$

Comparison with (1.27) and Box 1.1 shows that φ_{kn} are the components of a rotation about **u** = 0, i.e., the point P. The elements $\varphi_{kk} = 0$ and $\varphi_{kn} = -\varphi_{nk}$; the tensor is *antisymmetric* and its diagonal elements are all zero:

$$\varphi_{kn} = \begin{bmatrix} 0 & \varphi_{12} & \varphi_{13} \\ -\varphi_{12} & 0 & \varphi_{23} \\ -\varphi_{13} & -\varphi_{23} & 0 \end{bmatrix} \qquad (8.22)$$

The product of this tensor with the relative position vector y_n gives, in matrix form,

$$\varphi_{kn}y_n = \begin{bmatrix} 0 & \varphi_{12} & \varphi_{13} \\ -\varphi_{12} & 0 & \varphi_{23} \\ -\varphi_{13} & -\varphi_{23} & 0 \end{bmatrix}\begin{bmatrix} y_1 \\ y_2 \\ y_3 \end{bmatrix} = \begin{bmatrix} \varphi_{12}y_2 + \varphi_{13}y_3 \\ -\varphi_{12}y_1 + \varphi_{23}y_3 \\ -\varphi_{13}y_1 - \varphi_{23}y_2 \end{bmatrix} \qquad (8.23)$$

The column matrix on the right-hand side of this equation has the same components as the vector

$$\begin{vmatrix} \mathbf{e}_1 & \mathbf{e}_2 & \mathbf{e}_3 \\ -\varphi_{23} & \varphi_{13} & -\varphi_{12} \\ y_1 & y_2 & y_3 \end{vmatrix} = \boldsymbol{\varphi} \times \mathbf{y} \qquad (8.24)$$

Here $\mathbf{e}_1$, $\mathbf{e}_2$, and $\mathbf{e}_3$ are unit vectors for the x_1-, x_2-, and x_3-axes, respectively. The vector $\boldsymbol{\varphi}$ represents a rotation, while **y** denotes the position of an arbitrary point Q of the body relative to the point P, so $\boldsymbol{\varphi} \times \mathbf{y}$ describes an infinitesimal rigid-body *rotation* of the body about an axis through P. The direction of the rotation axis is the vector $\boldsymbol{\varphi}$ with components $(-\varphi_{23}, \varphi_{13}, -\varphi_{12})$. Following (8.21), this can also be written

$$\boldsymbol{\varphi} = \left(\frac{\partial u_3}{\partial x_2} - \frac{\partial u_2}{\partial x_3}\right)\mathbf{e}_1 + \left(\frac{\partial u_1}{\partial x_3} - \frac{\partial u_3}{\partial x_1}\right)\mathbf{e}_2 + \left(\frac{\partial u_2}{\partial x_1} - \frac{\partial u_1}{\partial x_2}\right)\mathbf{e}_3 \qquad (8.25)$$

$$\varphi = \begin{vmatrix} \mathbf{e}_1 & \mathbf{e}_2 & \mathbf{e}_3 \\ \partial/\partial x_1 & \partial/\partial x_2 & \partial/\partial x_3 \\ u_1 & u_2 & u_3 \end{vmatrix} = \nabla \times \mathbf{u} \qquad (8.26)$$

The rigid-body rotation is a displacement of the entire body without deformation. Neither the translation $\mathbf{u}$ nor the rotation φ of the rigid body takes part in the propagation of seismic waves.

The quantity ε_{kn} in (8.20) is the *strain* tensor. It describes a deformation in which different parts of the body are displaced relative to each other. As long as these displacements are small, the deformation is elastic and the strains can be described by a (3×3) strain matrix, whose general term is defined by (8.19):

$$\varepsilon_{kn} = \frac{1}{2} \left(\frac{\partial u_k}{\partial x_n} + \frac{\partial u_n}{\partial x_k} \right) \qquad (8.27)$$

It is evident from this definition that interchanging the indices does not change the general term; i.e., the strain matrix is *symmetric* ($\varepsilon_{kn} = \varepsilon_{nk}$). The diagonal terms of the strain matrix (i.e., ε_{kk}) describe *normal strains*, which correspond to changes in elongation of the body; the non-diagonal terms describe *shear strains*, which arise from angular distortion of the body.

8.3.1 Normal strain

Consider two points of a body that lie close to each other at the positions x_1 and $(x_1 + \delta x_1)$, respectively (Fig. 8.7(a)). If the body is stretched in the direction of the x_1-axis (Fig. 8.7(b)), the points are displaced by the small amounts u_1 and $(u_1 + \delta u_1)$, respectively. Using a MacLaurin or Taylor series, we can write

$$(u_1 + \delta u_1) = u_1 + \frac{\partial u_1}{\partial x_1} \delta x_1 + \frac{1}{2} \frac{\partial^2 u_1}{\partial x_1^2} (\delta x_1)^2 + \cdots \qquad (8.28)$$

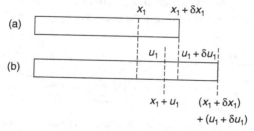

Fig. 8.7. Definition of normal strain for extension in the x_1-direction.

If the displacements are infinitesimally small, we can truncate the power series at first order, getting

$$\delta u_1 = \frac{\partial u_1}{\partial x_1} \delta x_1 \qquad (8.29)$$

The original separation of the two points was δx_1; after extension their separation is $(\delta x_1 + \delta u_1)$. The *normal strain* parallel to the x_1-axis is the fractional change in length resulting from an infinitesimal displacement parallel to the x_1-axis and is denoted ε_{11}; thus,

$$\varepsilon_{11} = \lim_{\delta x_1 \to 0} \frac{(\delta x_1 + \delta u_1) - \delta x_1}{\delta x_1} = \frac{\partial u_1}{\partial x_1} \qquad (8.30)$$

In a similar way, normal strains are defined for the x_2- and x_3-directions. If a point at x_k is displaced by an infinitesimal amount to $x_k + u_k$, then there arise normal strains ε_{kk}, corresponding to

$$\varepsilon_{kk} = \frac{\partial u_k}{\partial x_k} \qquad (8.31)$$

The normal strains are not independent of each other in an elastic body. Consider the change in shape of the bar in Fig. 8.8. When it is stretched parallel to the x_1-axis, it becomes thinner parallel to the x_2-axis and parallel to the x_3-axis. The transverse strains ε_{22} and ε_{33} are of opposite sign to the extension ε_{11}, but are proportional to it; so they can be expressed as

$$\frac{\varepsilon_{22}}{\varepsilon_{11}} = \frac{\varepsilon_{33}}{\varepsilon_{11}} = -v \qquad (8.32)$$

The constant of proportionality v is *Poisson's ratio*. The value of v is constrained to lie between 0 (no lateral contraction) and a maximum value of 0.5 for an incompressible fluid. In the Earth's interior, v has a value around 0.24–0.27. A body that has $v = 0.25$ is called an ideal Poisson body.

The normal strains result in a change of volume. The volume of the rectangular box in Fig. 8.5 is $V = \delta x_1\, \delta x_2\, \delta x_3$. As a result of infinitesimal displacements

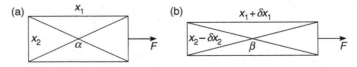

Fig. 8.8. Illustration of the lateral contraction and the change in the angles between the diagonals of a rectangular cross-section as a result of longitudinal extension.

δu_1, δu_2, and δu_3 the edges increase to $\delta x_1 + \delta u_1$, $\delta x_2 + \delta u_2$, and $\delta x_3 + \delta u_3$, respectively. The fractional change in volume is

$$\frac{\delta V}{V} = \frac{(\delta x_1 + \delta u_1)(\delta x_2 + \delta u_2)(\delta x_3 + \delta u_3) - \delta x_1 \, \delta x_2 \, \delta x_3}{\delta x_1 \, \delta x_2 \, \delta x_3}$$

$$= \left(\frac{\delta x_1 + \delta u_1}{\delta x_1}\right)\left(\frac{\delta x_2 + \delta u_2}{\delta x_2}\right)\left(\frac{\delta x_3 + \delta u_3}{\delta x_3}\right) - 1 \qquad (8.33)$$

The limit of the fractional change in volume, for small V, is defined as the dilatation, θ. As in (8.30) the limiting values of $\delta u_1/\delta x_1$, $\delta u_2/\delta x_2$, and $\delta u_3/\delta x_3$ are the longitudinal strains ε_{11}, ε_{22}, and ε_{33}, respectively. Thus

$$\theta = \lim_{V \to 0} \frac{\delta V}{V} = (1 + \varepsilon_{11})(1 + \varepsilon_{22})(1 + \varepsilon_{33}) - 1 \qquad (8.34)$$

This expression for θ contains second- and third-order products of the strains that can be neglected, thus

$$\theta = \varepsilon_{11} + \varepsilon_{22} + \varepsilon_{33} = \frac{\partial u_1}{\partial x_1} + \frac{\partial u_2}{\partial x_2} + \frac{\partial u_3}{\partial x_3} \qquad (8.35)$$

Taking **u** as the displacement vector, the dilatation θ is equivalent to

$$\theta = \nabla \cdot \mathbf{u} \qquad (8.36)$$

Using tensor notation, and the summation convention implied by a repeated index,

$$\theta = \varepsilon_{kk} = \frac{\partial u_k}{\partial x_k} \qquad (8.37)$$

8.3.2 Shear strain

The stress components (σ_{12}, σ_{23}, σ_{31}) act obliquely on the surface of the rectangular reference box (Fig. 8.4) and produce *shear strains*, which are manifested as changes in the angular relationships between parts of a body. These can also result from normal stresses. For example, the angles α and β between the internal diagonals of a rectangular cross-section (Fig. 8.8), before and after extension, respectively, are unequal; i.e., a longitudinal extension gives rise to shear strain as well as normal strain.

Consider the two-dimensional distortion of a rectangle $A_0B_0C_0D_0$ by shear stresses in the x_1–x_2 plane (Fig. 8.9). Point A_0 is displaced parallel to the x_1-axis by an amount u_1 and parallel to the x_2-axis by an amount u_2. The shear strain

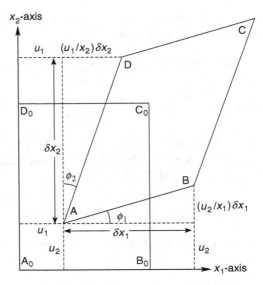

Fig. 8.9. Displacements accompanying two-dimensional shear strain in the x_1–x_2 plane.

causes point D_0, at a vertical distance δx_2 above A_0, to be displaced by an amount $(\partial u_1 / \partial x_2)\delta x_2$ parallel to the x_1-axis. This rotates side AD clockwise through a small angle ϕ_2. For infinitesimal displacements

$$\phi_2 = \tan \phi_2 = \frac{(\partial u_1 / \partial x_2)\delta x_2}{\delta x_2} = \frac{\partial u_1}{\partial x_2} \tag{8.38}$$

Similarly, point B_0, which is initially at a horizontal distance δx from A_0, is displaced by the amount $(\partial u_2 / \partial x_1)\delta x_1$ parallel to the x_2-axis, causing AB to rotate counterclockwise through a small angle ϕ_1 given by

$$\phi_1 = \tan \phi_1 = \frac{(\partial u_2 / \partial x_1)\delta x_1}{\delta x_1} = \frac{\partial u_2}{\partial x_1} \tag{8.39}$$

The shear-strain component ε_{12} is defined in (8.27):

$$\varepsilon_{12} = \frac{1}{2}\left(\frac{\partial u_2}{\partial x_1} + \frac{\partial u_1}{\partial x_2}\right) = \frac{1}{2}(\phi_1 + \phi_2) \tag{8.40}$$

Transposition of the indices 1 and 2 yields the shear-strain component ε_{21}, which is identical to ε_{12}. The total distortion in the x_1–x_2 plane is

$$\phi_1 + \phi_2 = \varepsilon_{12} + \varepsilon_{21} = 2\varepsilon_{12} = 2\varepsilon_{21} \tag{8.41}$$

The same argument leads to the definition of strain components ε_{23} ($=\varepsilon_{32}$) and ε_{31} ($=\varepsilon_{13}$) for angular distortions in the x_2–x_3 and x_3–x_1 planes, respectively. The shear strains are therefore

$$\varepsilon_{12} = \varepsilon_{21} = \frac{1}{2}\left(\frac{\partial u_2}{\partial x_1} + \frac{\partial u_1}{\partial x_2}\right)$$

$$\varepsilon_{23} = \varepsilon_{32} = \frac{1}{2}\left(\frac{\partial u_3}{\partial x_2} + \frac{\partial u_2}{\partial x_3}\right) \tag{8.42}$$

$$\varepsilon_{31} = \varepsilon_{13} = \frac{1}{2}\left(\frac{\partial u_1}{\partial x_3} + \frac{\partial u_3}{\partial x_1}\right)$$

They are expressed in tensor form by

$$\varepsilon_{kn} = \varepsilon_{nk} = \frac{1}{2}\left(\frac{\partial u_n}{\partial x_k} + \frac{\partial u_k}{\partial x_n}\right) \tag{8.43}$$

The longitudinal and shear strains together form the symmetric *strain matrix*

$$\varepsilon_{kn} = \begin{pmatrix} \varepsilon_{11} & \varepsilon_{12} & \varepsilon_{13} \\ \varepsilon_{21} & \varepsilon_{22} & \varepsilon_{23} \\ \varepsilon_{31} & \varepsilon_{32} & \varepsilon_{33} \end{pmatrix} \tag{8.44}$$

The elements of the matrix represent the strain tensor ε_{kn} ($k = 1, 2, 3$; $n = 1, 2, 3$), which, because of its symmetry, has six independent elements.

8.4 Perfectly elastic stress–strain relationships

Hooke's law describes perfectly elastic deformation, which occurs by means of infinitesimal strains. The components of strain are then linear functions of the components of stress. The linear dependence allows the definition of elastic moduli, each of which is a constant of proportionality between stress and strain. Young's modulus, the shear modulus, and the bulk modulus relate the different elements of the stress and strain tensors for appropriate types of deformation.

Young's modulus

Each normal stress σ_{kk} is proportional to the corresponding normal strain ε_{kk}. Thus,

$$\sigma_{kk} = E\varepsilon_{kk} \tag{8.45}$$

The constant of proportionality, E, is *Young's modulus*. The lateral contraction that accompanies longitudinal extension is described by Poisson's ratio, v (see (8.32)).

Shear modulus (or rigidity modulus)

The shear strain ε_{kn} (i.e., the total angular distortion) in a plane is proportional to the shear stress in the plane, σ_{kn}. Equation (8.41) defines the shear strain, so for $k \neq n$ we have the relationship

$$\sigma_{kn} = 2\mu\varepsilon_{kn} \qquad (8.46)$$

The constant of proportionality, μ, is the *rigidity* (or *shear*) *modulus*.

Bulk modulus (or incompressibility)

The *bulk modulus*, K, is a measure of the change of pressure needed to cause a change of volume. A body under hydrostatic pressure p (defined as acting inwards, equivalent to a negative normal stress) experiences a change of volume. The fractional change in volume is the dilatation, θ, which is related to the principal strains as in (8.34)–(8.37). Under hydrostatic conditions there are no shear stresses ($\sigma_{kn} = 0$) and the normal stresses are equal ($\sigma_{kk} = -p$). The dilatation is proportional to the pressure and the constant of proportionality is K. Thus, we have the simple relationships

$$p = -K\theta = -K\frac{\partial u_k}{\partial x_k} = -K \nabla \cdot \mathbf{u} \qquad (8.47)$$

8.4.1 The Lamé constants

A change of length in the x_1-direction consists of the extension due to σ_{11} and contributions from the lateral contractions in the x_2- and x_3-directions that are due to σ_{22} and σ_{33}. The normal strain equals σ_{11}/E and, using (8.32), the lateral contractions contribute $-v\sigma_{22}/E$ and $-v\sigma_{33}/E$, respectively, to the longitudinal strain. Thus, for the x_1-direction

$$\varepsilon_{11} = \frac{\sigma_{11}}{E} - v\frac{\sigma_{22}}{E} - v\frac{\sigma_{33}}{E} \qquad (8.48)$$

Similar equations are obtained for the x_2- and x_3-directions. On multiplying each equation throughout by E, we get the set of equations

$$E\varepsilon_{11} = \sigma_{11} - v\sigma_{22} - v\sigma_{33}$$
$$E\varepsilon_{22} = \sigma_{22} - v\sigma_{33} - v\sigma_{11} \qquad (8.49)$$
$$E\varepsilon_{33} = \sigma_{33} - v\sigma_{11} - v\sigma_{22}$$

Adding these equations gives

$$E(\varepsilon_{11} + \varepsilon_{22} + \varepsilon_{33}) = (\sigma_{11} + \sigma_{22} + \sigma_{33})(1 - 2v) \qquad (8.50)$$

$$E\theta = (\sigma_{11} + \sigma_{22} + \sigma_{33})(1 - 2v) \qquad (8.51)$$

This equation can be rewritten for σ_{11}:

$$\sigma_{11} = \frac{E}{1 - 2v}\theta - (\sigma_{22} + \sigma_{33}) \tag{8.52}$$

We can obtain another expression for the sum $(\sigma_{22} + \sigma_{33})$ from the first line of (8.49):

$$(\sigma_{22} + \sigma_{33}) = -\frac{E\varepsilon_{11} - \sigma_{11}}{v} \tag{8.53}$$

Substituting this expression for $(\sigma_{22} + \sigma_{33})$ into (8.52) gives

$$\sigma_{11} = \frac{E}{1 - 2v}\theta + \frac{E\varepsilon_{11} - \sigma_{11}}{v} \tag{8.54}$$

$$v\sigma_{11} = \frac{Ev}{1 - 2v}\theta + E\varepsilon_{11} - \sigma_{11} \tag{8.55}$$

$$\sigma_{11} = \frac{Ev}{(1 - 2v)(1 + v)}\theta + \frac{E}{1 + v}\varepsilon_{11} \tag{8.56}$$

The coefficients of θ and ε_{11} define the *Lamé constants* λ and μ, respectively:

$$\lambda = \frac{Ev}{(1 - 2v)(1 + v)} \tag{8.57}$$

$$2\mu = \frac{E}{1 + v} \tag{8.58}$$

The relationship between normal stress and normal strain in terms of the Lamé constants is

$$\sigma_{11} = \lambda\theta + 2\mu\varepsilon_{11} \tag{8.59}$$

A similar result would be obtained by using any line in (8.49), so in general the normal stresses and strains are related by

$$\sigma_{kk} = \lambda\theta + 2\mu\varepsilon_{kk} \tag{8.60}$$

The Lamé constant μ is equivalent to the shear modulus. This can be shown by establishing independently the relationship among Young's modulus, the shear modulus, and Poisson's ratio (Box. 8.1), which leads to the same equation as that in (8.58). The shear modulus is defined in (8.46) as the ratio of the shear stress σ_{kn} to the shear strain ε_{kn}. Using the Kronecker-delta symbol, we can therefore write the more general relationship

$$\sigma_{kn} = \lambda\theta\delta_{kn} + 2\mu\varepsilon_{kn} \tag{8.61}$$

Box 8.1.   **Relationship of the shear modulus, Young's modulus, and Poisson's ratio**

Consider a body with a square cross-section subject to normal stresses in the x_1–x_2 plane only (i.e., $\sigma_{33} = 0$), as in Fig. B8.1.1(a). Let the area of each side normal to the figure be A. Let p be the average of the normal stresses σ_{11} and σ_{22}, and let σ be the stress difference between p and each normal stress. Therefore

$$\sigma = \sigma_{11} - p = p - \sigma_{22} \tag{1}$$

The outward stress difference σ along the x_1-axis causes extension, whereas the inward stress difference $-\sigma$ along the x_2-axis causes contraction (Fig. B8.1.1(b)). The change of shape of the cross-section results in angular distortions internally. Thus the normal stresses give rise to both normal strains and shear strains.

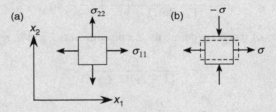

Fig. B8.1.1. (a) Normal stresses σ_{11} and σ_{22} in the x_1–x_2 plane. (b) Deviatoric stresses $\pm \sigma$, equal to the difference between the normal stresses and their mean value.

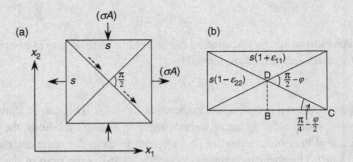

Fig. B8.1.2. (a) Undeformed square cross-section showing inward and outward forces (σA) due to deviatoric stresses. (b) Side lengths, normal strains, and changes to the angles between intersecting diagonals as a result of deviatoric stresses.

The outward force in the x_1-direction is (σA), which has a component $(\sigma A)/\sqrt{2}$ along the body diagonal (Fig. B8.1.2(a)). Likewise, the inward force in the x_2-direction has a component $(\sigma A)/\sqrt{2}$ in the same direction. The combined force parallel to the diagonal is $\sqrt{2}(\sigma A)$. The area of a side normal to the cross-section is A, so the area of a normal planar section that includes the diagonal is $\sqrt{2}A$. The *shear stress* along the diagonal is therefore equal to σ.

The diagonals are initially at right angles to each other, but after deformation their mutual orientation changes by an angle φ (Fig B8.1.2(b)), which, as defined in Section 8.3.2, is the shear strain in the x_1–x_2 plane. Consider the angles and side lengths in the triangle BCD. If the original side length of the square cross-section is s (Fig. B8.1.2(a)), the side along the x-axis extends to $s(1 + \varepsilon_{11})$ while the side normal to this contracts to $s(1 + \varepsilon_{22})$. The tangent of the angle BCD is DB/BC; thus,

$$\tan\left(\frac{\pi}{4} - \frac{\varphi}{2}\right) = \frac{s(1 + \varepsilon_{22})/2}{s(1 + \varepsilon_{11})/2} = \frac{1 + \varepsilon_{22}}{1 + \varepsilon_{11}} \tag{2}$$

The trigonometric formula for the tangent of the difference of two angles gives

$$\tan\left(\frac{\pi}{4} - \frac{\varphi}{2}\right) = \frac{\tan \pi/4 - \tan \varphi/2}{1 + \tan \pi/4 \tan \varphi/2} = \frac{1 - \tan \varphi/2}{1 + \tan \varphi/2} \tag{3}$$

On equating the two expressions, we have

$$\frac{1 + \varepsilon_{22}}{1 + \varepsilon_{11}} = \frac{1 - \tan \varphi/2}{1 + \tan \varphi/2} \tag{4}$$

From (8.46), with $\sigma_{33} = 0$ and replacing the normal stresses by the deforming stress differences, we can write expressions for ε_{11} and ε_{22},

$$\varepsilon_{11} = \frac{\sigma_{11}}{E} - v\frac{\sigma_{22}}{E} = \frac{\sigma}{E} - v\frac{(-\sigma)}{E} = \frac{\sigma}{E}(1 + v) \tag{5}$$

$$\varepsilon_{22} = \frac{\sigma_{22}}{E} - v\frac{\sigma_{11}}{E} = \frac{(-\sigma)}{E} - v\frac{\sigma}{E} = -\frac{\sigma}{E}(1 + v) \tag{6}$$

We now insert these expressions into (4). Note that the angle φ is very small, so we can replace the tangent of the angle by the angle itself,

$$\frac{1 - \sigma/E(1 + v)}{1 + \sigma/E(1 + v)} = \frac{1 - \varphi/2}{1 + \varphi/2} \tag{7}$$

$$\frac{\varphi}{2} = \frac{\sigma}{E}(1 + v) \tag{8}$$

The shear modulus μ is the ratio of the shear stress to the shear strain; in this case, the ratio of the deforming stress σ to the angular distortion φ:

$$\mu = \frac{\sigma}{\varphi} \tag{9}$$

From (8) we therefore have the following relationship among the shear modulus μ, Young's modulus E, and Poisson's ratio v:

$$\mu = \frac{E}{2(1 + v)} \tag{10}$$

8.5 The seismic wave equation

In order to describe the propagation of a seismic wave in the Earth, some simplifying assumptions must be made. First, the heterogeneity of the medium is neglected. We assume that the medium is uniform and isotropic. This allows us to use the homogeneous equation of motion derived in (8.16) to describe particle displacements. Secondly, the medium is assumed to behave as a perfectly elastic substance; only infinitesimal displacements of the particles of the medium are considered. The relationship between stress and strain is governed by (8.61). The equation of motion becomes

$$\rho \frac{\partial^2 u_n}{\partial t^2} = \frac{\partial}{\partial x_k}(\lambda \theta \delta_{kn} + 2\mu \varepsilon_{kn}) \tag{8.62}$$

Next we assume that the Lamé parameters λ and μ do not vary with position, and therefore can be treated as constants. This implies in effect that there are no velocity gradients in the medium. On writing $\theta = \varepsilon_{nn}$ and observing the Kronecker delta, we have

$$\rho \frac{\partial^2 u_n}{\partial t^2} = \lambda \frac{\partial \varepsilon_{nn}}{\partial x_n} + 2\mu \frac{\partial \varepsilon_{kn}}{\partial x_k} \tag{8.63}$$

Now we can insert the definitions of ε_{nn} from (8.37) and ε_{kn} from (8.43),

$$\rho \frac{\partial^2 u_n}{\partial t^2} = \lambda \frac{\partial}{\partial x_n}\left(\frac{\partial u_k}{\partial x_k}\right) + \mu \frac{\partial}{\partial x_k}\left(\frac{\partial u_n}{\partial x_k} + \frac{\partial u_k}{\partial x_n}\right) \tag{8.64}$$

$$\rho \frac{\partial^2 u_n}{\partial t^2} = \lambda \frac{\partial}{\partial x_n} \frac{\partial u_k}{\partial x_k} + \mu \frac{\partial^2 u_n}{\partial x_k^2} + \mu \frac{\partial}{\partial x_k} \frac{\partial u_k}{\partial x_n} \tag{8.65}$$

Note that the order of differentiation in the last term can be interchanged without altering the meaning:

$$\frac{\partial}{\partial x_k} \frac{\partial u_k}{\partial x_n} = \frac{\partial^2 u_k}{\partial x_k \partial x_n} = \frac{\partial}{\partial x_n} \frac{\partial u_k}{\partial x_k} \tag{8.66}$$

After gathering terms and simplifying, we have

$$\rho \frac{\partial^2 u_n}{\partial t^2} = (\lambda + \mu) \frac{\partial}{\partial x_k} \left(\frac{\partial u_k}{\partial x_n} \right) + \mu \frac{\partial^2 u_n}{\partial x_k^2} \tag{8.67}$$

In symbolic form this equation is

$$\rho \frac{\partial^2 \mathbf{u}}{\partial t^2} = (\lambda + \mu) \nabla (\nabla \cdot \mathbf{u}) + \mu \nabla^2 \mathbf{u} \tag{8.68}$$

Now we recall the vector identity in (1.34) to obtain an expression for $\nabla^2 \mathbf{u}$:

$$\nabla^2 \mathbf{u} = \nabla (\nabla \cdot \mathbf{u}) - \nabla \times (\nabla \times \mathbf{u}) \tag{8.69}$$

The homogeneous equation of motion becomes

$$\rho \frac{\partial^2 \mathbf{u}}{\partial t^2} = (\lambda + \mu) \nabla (\nabla \cdot \mathbf{u}) + \mu (\nabla (\nabla \cdot \mathbf{u}) - \nabla \times (\nabla \times \mathbf{u})) \tag{8.70}$$

$$\rho \frac{\partial^2 \mathbf{u}}{\partial t^2} = (\lambda + 2\mu) \nabla (\nabla \cdot \mathbf{u}) - \mu (\nabla \times (\nabla \times \mathbf{u})) \tag{8.71}$$

This is the starting point for the treatment of elastic waves in an isotropic homogeneous medium.

Minerals are individually anisotropic, their properties being controlled by their crystal structure. However, in a large enough assemblage, random ordering of the crystals makes a material macroscopically isotropic and justifies the assumption of this condition for the Earth's interior. The assumption of homogeneity is unrealistic. For example, the density and elastic parameters that control the passage of seismic disturbances change with depth and may also vary laterally at a given depth. However, a heterogeneous medium can be modeled acceptably by dividing it into smaller elements (e.g., parallel horizontal layers, or small blocks) and assuming homogeneous conditions in each element. Real conditions can then be approximated by judicious choice of the thickness, density, and elastic parameters of each element.

The assumption that seismic signals propagate by elastic displacements of the medium is true only at some distance from the source. In an earthquake or explosion the medium immediately surrounding the source is destroyed, particle displacements are large and permanent, and the deformation is anelastic. However, the elastic conditions underlying (8.71) are applicable for the passage of a seismic disturbance at a distance from its source.

In order to proceed further with the equations of motion for seismic body waves we take separately the divergence and curl of both sides of (8.71). This leads to the description of primary and secondary seismic waves.

8.5.1 Primary waves (P-waves)

First we take the divergence of both sides of (8.71):

$$\rho \frac{\partial^2 (\nabla \cdot \mathbf{u})}{\partial t^2} = (\lambda + 2\mu)\nabla \cdot \nabla(\nabla \cdot \mathbf{u}) - \mu(\nabla \cdot \nabla \times (\nabla \times \mathbf{u})) \tag{8.72}$$

The vector identity (1.33) states that the divergence of the curl of any vector $\mathbf{a}$ is zero, i.e., $\nabla \cdot (\nabla \times \mathbf{a}) = 0$. Thus the second term on the right is zero, and we get

$$\rho \frac{\partial^2 (\nabla \cdot \mathbf{u})}{\partial t^2} = (\lambda + 2\mu)\nabla^2(\nabla \cdot \mathbf{u}) \tag{8.73}$$

The dilatation θ, defined as the fractional change in volume, was shown in (8.36) to equal the divergence of the displacement vector $\mathbf{u}$; thus,

$$\rho \frac{\partial^2 \theta}{\partial t^2} = (\lambda + 2\mu)\nabla^2 \theta \tag{8.74}$$

$$\frac{\partial^2 \theta}{\partial t^2} = \alpha^2 \nabla^2 \theta \tag{8.75}$$

where

$$\alpha^2 = \frac{\lambda + 2\mu}{\rho} \tag{8.76}$$

On examining both sides of (8.75) it is evident that α has the dimensions of a velocity. It is the velocity with which a change in volume (dilatation) propagates through the medium. The disturbance propagates as a succession of compressions and dilatations with velocity α. The corresponding seismic wave is the *primary wave*, or *P-wave*, so called because it is the first arrival on the record of a seismic event.

The bulk modulus, Young's modulus, and Poisson's ratio can each be expressed solely in terms of the Lamé constants (Box 8.2). The relationship between the bulk modulus and the Lamé constants allows us to write (8.76) as

Box 8.2. **Elastic parameters and the Lamé constants**

1. **The bulk modulus, K**

The bulk modulus describes volumetric shape changes of a material under the effects of the *normal* stresses σ_{11}, σ_{22}, and σ_{33}. Hooke's law for each normal stress gives the equations

$$\begin{aligned}
\sigma_{11} &= \lambda\theta + 2\mu\varepsilon_{11} \\
\sigma_{22} &= \lambda\theta + 2\mu\varepsilon_{22} \\
\sigma_{33} &= \lambda\theta + 2\mu\varepsilon_{33}
\end{aligned} \tag{1}$$

Adding these equations together gives

$$\sigma_{11} + \sigma_{22} + \sigma_{33} = 3\lambda\theta + 2\mu(\varepsilon_{11} + \varepsilon_{22} + \varepsilon_{33}) \tag{2}$$

The dilatation θ is defined as

$$\theta = \varepsilon_{11} + \varepsilon_{22} + \varepsilon_{33} \tag{3}$$

For hydrostatic conditions $\sigma_{11} = \sigma_{22} = \sigma_{33} = -p$. Substituting into (2) and re-arranging gives

$$-3p = 3\lambda\theta + 2\mu\theta \tag{4}$$

The definition of the bulk modulus is $K = -p/\theta$. Therefore,

$$K = \lambda + \frac{2}{3}\mu \tag{5}$$

2. **Young's modulus, E**

When a uniaxial normal stress is applied to a material, there results a longitudinal extension or shortening that is proportional to the stress. The constant of proportionality is Young's modulus. Suppose that the applied stress is along the x_1-axis, so that $\sigma_{yy} = \sigma_{zz} = 0$. Hooke's law applied to each axis gives

$$\begin{aligned}
\sigma_{11} &= \lambda\theta + 2\mu\varepsilon_{11} \\
0 &= \lambda\theta + 2\mu\varepsilon_{22} \\
0 &= \lambda\theta + 2\mu\varepsilon_{33}
\end{aligned} \tag{6}$$

Adding both sides of these equations gives

$$\sigma_{11} = 3\lambda\theta + 2\mu(\varepsilon_{11} + \varepsilon_{22} + \varepsilon_{33}) = 3\lambda\theta + 2\mu\theta \tag{7}$$

$$\theta = \frac{\sigma_{11}}{(3\lambda + 2\mu)} \tag{8}$$

Inserting this into the first line of (6) gives

$$\sigma_{11} = \lambda\frac{\sigma_{11}}{3\lambda + 2\mu} + 2\mu\varepsilon_{11} \tag{9}$$

After gathering and rearranging terms,

$$\sigma_{11}\left(1 - \frac{\lambda}{3\lambda + 2\mu}\right) = 2\mu\varepsilon_{11} \tag{10}$$

$$\sigma_{11} = \mu\left(\frac{3\lambda + 2\mu}{\lambda + \mu}\right)\varepsilon_{11} \tag{11}$$

The definition of Young's modulus is $E = \sigma_{11}/\varepsilon_{11}$, so

$$E = \mu\left(\frac{3\lambda + 2\mu}{\lambda + \mu}\right) \tag{12}$$

3.   Poisson's ratio, v

The definitions of the Lamé constants in (8.57) and (8.58) give, respectively,

$$\lambda = \frac{Ev}{(1 - 2v)(1 + v)} = \frac{E}{(1 + v)}\frac{v}{(1 - 2v)} \tag{13}$$

$$2\mu = \frac{E}{1 + v} \tag{14}$$

On combining these equations we obtain

$$\lambda = 2\mu\frac{v}{1 - 2v} \tag{15}$$

In terms of the Lamé constants, Poisson's ratio v is given by

$$v = \frac{\lambda}{2(\lambda + \mu)} \tag{16}$$

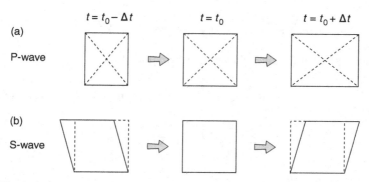

Fig. 8.10. Schematic illustration of (a) changes of volume and the angles between intersecting diagonals during passage of a P-wave, and (b) the change of shape due to shear during passage of an S-wave.

$$\alpha^2 = \frac{1}{\rho}\left(\lambda + \frac{2}{3}\mu + \frac{4}{3}\mu\right) = \frac{1}{\rho}\left(K + \frac{4}{3}\mu\right) \qquad (8.77)$$

The velocity of the P-wave depends both on the bulk modulus (or incompressibility) and on the shear modulus. Thus a P-wave can propagate through a fluid phase in which the shear modulus μ is zero.

The propagation of a one-dimensional compression is illustrated in Fig. 8.10(a), which shows an undeformed volume at time t_0, the compressed volume at an earlier time $t_0 - \Delta t$, and the dilated volume at a later time $t_0 + \Delta t$. Changes of the angles between the diagonals of the original square demonstrate that the deformation in the compressional wave also has a shearing aspect.

8.5.2 Secondary waves (S-waves)

Next, we proceed to take the curl of both sides of (8.71):

$$\rho\frac{\partial^2(\nabla \times \mathbf{u})}{\partial t^2} = (\lambda + 2\mu)(\nabla \times \nabla(\nabla \cdot \mathbf{u})) - \mu(\nabla \times \nabla \times (\nabla \times \mathbf{u})) \qquad (8.78)$$

Again we use a vector identity to simplify the equation. The identity in (1.32) states that the curl of the gradient of any scalar function f is zero, i.e., $\nabla \times \nabla f = 0$. Thus the first term on the right is zero. The remaining equation is

$$\rho\frac{\partial^2(\nabla \times \mathbf{u})}{\partial t^2} = -\mu(\nabla \times \nabla \times (\nabla \times \mathbf{u})) \qquad (8.79)$$

Now we again use the vector identity (1.34), obtaining

$$\rho \frac{\partial^2 (\nabla \times \mathbf{u})}{\partial t^2} = -\mu \, \nabla(\nabla \cdot (\nabla \times \mathbf{u})) + \mu \, \nabla^2(\nabla \times \mathbf{u}) \qquad (8.80)$$

The divergence of the curl of a vector is zero, therefore

$$\rho \frac{\partial^2 (\nabla \times \mathbf{u})}{\partial t^2} = \mu \, \nabla^2(\nabla \times \mathbf{u}) \qquad (8.81)$$

$$\frac{\partial^2 (\nabla \times \mathbf{u})}{\partial t^2} = \beta^2 \, \nabla^2(\nabla \times \mathbf{u}) \qquad (8.82)$$

where

$$\beta^2 = \frac{\mu}{\rho} \qquad (8.83)$$

The components of $\nabla \times \mathbf{u}$ are in the plane normal to the displacement $\mathbf{u}$. The disturbance propagates through the medium as a succession of shear displacements and travels with velocity β. Because it depends on the shear modulus, which is zero in liquids and gases, a shear wave can propagate only in solid materials.

Comparison of (8.77) and (8.83) yields the seismic parameter Φ, defined as

$$\Phi = \alpha^2 - \frac{4}{3}\beta^2 = \frac{K}{\rho} \qquad (8.84)$$

This parameter is important for determining the variation of density as well as the adiabatic temperature gradient inside the Earth, which can be computed because the P-wave and S-wave velocities are well known as functions of depth.

The S-wave velocity β is less than the P-wave velocity α. As a result the seismic *shear wave* (or *S-wave*) is recorded at a seismic station later than the P-wave, so it is also called the *secondary wave*. During the propagation of a one-dimensional shear deformation (Fig. 8.10(b)), the shape of an originally square cross-section at time t_0 is distorted to a parallelogram at times $t_0 - \Delta t$ and $t_0 + \Delta t$. The area of the parallelogram is, however, the same as that of the original square. In three dimensions the shear wave propagates without change in volume.

8.5.3 Displacement potentials

A theorem established by Helmholtz shows that a vector field such as the displacement vector $\mathbf{u}$ can be expressed in terms of both a scalar potential φ and a vector potential $\boldsymbol{\psi}$, provided that the scalar field is irrotational ($\nabla \times \varphi = 0$) and the vector field is divergence-free ($\nabla \cdot \boldsymbol{\psi} = 0$). Thus

$$\mathbf{u} = \nabla\varphi + \nabla \times \boldsymbol{\psi} \tag{8.85}$$

An irrotational displacement field is one that has no shear components, whereas a divergence-free displacement takes place without change of volume. Consequently, in a seismic disturbance the potentials φ and $\boldsymbol{\psi}$ correspond to the displacements in P- and S-waves, respectively, and are obtained by solving the corresponding wave equations.

P-waves

On taking the divergence of $\mathbf{u}$ and noting that $\nabla \cdot (\nabla \times \boldsymbol{\psi}) = 0$, we have

$$\nabla \cdot \mathbf{u} = \nabla^2 \varphi \tag{8.86}$$

Substituting into (8.73) with α as the P-wave velocity gives

$$\frac{\partial^2 \left(\nabla^2 \varphi\right)}{\partial t^2} = \alpha^2 \, \nabla^2 \left(\nabla^2 \varphi\right) \tag{8.87}$$

$$\nabla^2 \left[\frac{\partial^2 \varphi}{\partial t^2} - \alpha^2 \, \nabla^2 \varphi \right] = 0 \tag{8.88}$$

This equation is always true if the expression in square brackets is zero. The defining equation for the scalar potential φ of the P-wave displacement is therefore

$$\frac{\partial^2 \varphi}{\partial t^2} - \alpha^2 \, \nabla^2 \varphi = 0 \tag{8.89}$$

S-waves

Next, taking the curl of $\mathbf{u}$, we have

$$\nabla \times \mathbf{u} = (\nabla \times \nabla\varphi) + (\nabla \times \nabla \times \boldsymbol{\psi}) \tag{8.90}$$

Using the identities in (1.32) and (1.34), we get

$$\nabla \times \mathbf{u} = \nabla(\nabla \cdot \boldsymbol{\psi}) - \nabla^2 \boldsymbol{\psi} \tag{8.91}$$

On applying the condition that the vector potential be divergence-free ($\nabla \cdot \boldsymbol{\psi} = 0$), this becomes

$$\nabla \times \mathbf{u} = -\nabla^2 \boldsymbol{\psi} \tag{8.92}$$

Substituting into (8.82) with β as the S-wave velocity gives

$$\frac{\partial^2 (\nabla^2 \psi)}{\partial t^2} = \beta^2 \, \nabla^2 (\nabla^2 \psi) \tag{8.93}$$

$$\nabla^2 \left[\frac{\partial^2 \psi}{\partial t^2} - \beta^2 \, \nabla^2 \psi \right] = 0 \tag{8.94}$$

Here again the equation is true if the expression in square brackets is zero. This leads to a defining equation for the vector potential ψ of the S-wave displacement:

$$\frac{\partial^2 \psi}{\partial t^2} - \beta^2 \, \nabla^2 \psi = 0 \tag{8.95}$$

8.6 Solutions of the wave equation

The *wavefront* of a seismic wave is defined as a surface in which all particles vibrate in phase with each other. Close to a point source in a homogeneous medium, the wavefronts form spheres around the source, and the wave is called a *spherical wave*. With increasing distance from the source the curvature of the spherical wavefront decreases and eventually becomes flat enough to be regarded as a plane. The normal to the wavefront is the direction of propagation of the wave, called the seismic *ray path*. Far from its source a seismic wave is called a *plane wave* and it may be described using orthogonal Cartesian coordinates.

8.6.1 One-dimensional solution for plane P-waves

For a plane P-wave propagating in the x_1-direction the x_2- and x_3-axes are perpendicular to each other in the plane of the wavefront. There is no change in the x_2- and x_3-directions, so derivatives with respect to these coordinates are zero. Equation (8.89) can then be written

$$\frac{1}{\alpha^2} \frac{\partial^2 \varphi}{\partial t^2} = \frac{\partial^2 \varphi}{\partial x_1^2} \tag{8.96}$$

In this equation φ is a function of both time and position. Invoking the method of separation of variables, we can write

$$\varphi(x_1, t) = X(x_1) T(t) \tag{8.97}$$

Upon inserting this into the equation and dividing both sides by φ we get

$$\frac{1}{\alpha^2 T}\frac{\partial^2 T}{\partial t^2} = \frac{1}{X}\frac{\partial^2 X}{\partial x_1^2} = -k_\alpha^2 \qquad (8.98)$$

Each side is a function of only one variable, so each side must equal the same constant, which we write as $-k_\alpha^2$. The negative sign is chosen so as to deliver periodic solutions. We get the equations

$$\frac{1}{\alpha^2 T}\frac{\partial^2 T}{\partial t^2} = -k_\alpha^2$$
$$\frac{1}{X}\frac{\partial^2 X}{\partial x_1^2} = -k_\alpha^2 \qquad (8.99)$$

Rearranging the equations gives

$$\frac{\partial^2 T}{\partial t^2} + k_\alpha^2 \alpha^2 T = 0$$
$$\frac{\partial^2 X}{\partial x_1^2} + k_\alpha^2 X = 0 \qquad (8.100)$$

These are simple harmonic motions. If we define $\omega = k_\alpha \alpha$, the separate solutions for the dependence on time and position are

$$T = T_1 \exp(i\omega t) + T_2 \exp(-i\omega t)$$
$$X = X_1 \exp(ik_\alpha x_1) + X_2 \exp(-ik_\alpha x_1) \qquad (8.101)$$

k_α is called the wave-number and ω the angular frequency of the P-wave. The general solution for a P-wave traveling along the x_1-axis is obtained by combining the partial solutions:

$$\varphi(x_1, t) = A \exp[i(\omega t + k_\alpha x_1)] + B \exp[-i(\omega t + k_\alpha x_1)]$$
$$+ C \exp[i(\omega t - k_\alpha x_1)] + D \exp[-i(\omega t - k_\alpha x_1)] \qquad (8.102)$$

The solution contains four arbitrary constants ($A = T_1 X_1$, $B = T_2 X_2$, $C = T_1 X_2$, $D = T_2 X_1$), whose values in a given situation are determined by the boundary conditions. If we consider only the real parts of the solutions (with new constants $A_1 = A + B$, $A_2 = C + D$), we obtain

$$\varphi(x_1, t) = A_1 \cos(\omega t + k_\alpha x_1) + A_2 \cos(\omega t - k_\alpha x_1) \qquad (8.103)$$

The two parts of the solution have phases $(\omega t + k_\alpha x_1)$ and $(\omega t - k_\alpha x_1)$, respectively. The velocity α with which a constant phase travels is called the *phase velocity*. The propagation of a constant phase of the first solution is governed by the condition that $(\omega t + k_\alpha x_1)$ is constant. On differentiating with respect to time, with ω and k_α held constant (and therefore also α, because $\alpha = \omega/k_\alpha$), we get

$$\frac{dx_1}{dt} = -\frac{\omega}{k_\alpha} = -\alpha \tag{8.104}$$

The negative sign indicates that this phase is a P-wave propagating with velocity α in the *negative* x_1-direction. The second part of the solution can be treated in the same way. It is seen to describe a P-wave propagating with velocity α in the *positive* x_1-direction. The velocity α is known as the phase velocity of the wave.

8.6.2 One-dimensional solution for plane S-waves

Using (8.95), the equation for the vector potential of an S-wave traveling in the direction of the x_1-axis can be written for each component ψ_n as

$$\frac{1}{\beta^2}\frac{\partial^2 \psi_n}{\partial t^2} = \frac{\partial^2 \psi_n}{\partial x_1^2} \tag{8.105}$$

This wave equation is solved as for P-waves, yielding solutions akin to (8.103). For S-waves propagating with velocity β, the wave-number is k_β and the components of the vector potential are

$$\psi_n(x_1, t) = B_{n1} \cos(\omega t + k_\beta x_1) + B_{n2} \cos(\omega t - k_\beta x_1) \tag{8.106}$$

The solutions describe shear waves that travel in the negative and positive x_1-directions with wave-number k_β and phase velocity $\beta = \omega/k_\beta$.

8.7 Three-dimensional propagation of plane P- and S-waves

The assumption that the plane wave is traveling along the x_1-axis is too restrictive. It is common usage in seismology (and other geophysical disciplines) to define Cartesian coordinates so that the vertical direction is the x_3-axis and the horizontal surface is the plane defined by the x_1- and x_2-axes. Box 8.3 shows how the one-dimensional solutions can be extended to three dimensions. This is applicable to both P-waves and S-waves. The solutions of the wave equation depend on the velocity of the wave, which determines the wave-number. For P-waves we have $|\mathbf{k}_\alpha| = \omega/\alpha$, and for S-waves $|\mathbf{k}_\beta| = \omega/\beta$.

Box 8.3. Three-dimensional solution of the wave equation

Let e_1, e_2, and e_3 be unit vectors corresponding to a set of Cartesian coordinates x_1, x_2, and x_3. The P-wave equation then becomes

$$\frac{1}{\alpha^2}\frac{\partial^2 \varphi}{\partial t^2} = \frac{\partial^2 \varphi}{\partial x_1^2} + \frac{\partial^2 \varphi}{\partial x_2^2} + \frac{\partial^2 \varphi}{\partial x_3^2} \tag{1}$$

and the solution by the method of separation of variables involves three spatial components,

$$\varphi(x_1, x_2, x_3, t) = X_1(x_1)X_2(x_2)X_3(x_3)T(t) \tag{2}$$

Inserting the solution and dividing throughout by φ, as for one-dimensional propagation, gives

$$\frac{1}{\alpha^2 T}\frac{\partial^2 T}{\partial t^2} = \frac{1}{X_1}\frac{\partial^2 X_1}{\partial x_1^2} + \frac{1}{X_2}\frac{\partial^2 X_2}{\partial x_2^2} + \frac{1}{X_3}\frac{\partial^2 X_3}{\partial x_3^2} = -k^2 \tag{3}$$

The constant $-k^2$ is equal to both the time-dependent and the spatially dependent parts of the solution. Continuing as for the one-dimensional case, by successively separating parts that depend on different coordinates on opposite sides of the equality sign, we get for the time-dependent variation

$$\frac{1}{\alpha^2 T}\frac{\partial^2 T}{\partial t^2} = -k^2 \tag{4}$$

This is a simple harmonic motion with angular frequency $\omega = k\alpha$. The solution is

$$T = T_0 \exp(\pm i\omega t) \tag{5}$$

The spatial variations are

$$\frac{1}{X_1}\frac{\partial^2 X_1}{\partial x_1^2} = -k^2 - \left(\frac{1}{X_2}\frac{\partial^2 X_2}{\partial x_2^2} + \frac{1}{X_3}\frac{\partial^2 X_3}{\partial x_3^2}\right) = -k_1^2 \tag{6}$$

$$\frac{1}{X_2}\frac{\partial^2 X_2}{\partial x_2^2} = -(k^2 - k_1^2) - \frac{1}{X_3}\frac{\partial^2 X_3}{\partial x_3^2} = -k_2^2 \tag{7}$$

$$\frac{1}{X_3}\frac{\partial^2 X_3}{\partial x_3^2} = -(k^2 - k_1^2 - k_2^2) = -k_3^2 \tag{8}$$

Positive and negative values of k_1, k_2, k_3, and ω satisfy these equations. We choose a particular solution that corresponds to a wave traveling in the direction of the positive reference axes:

$$\varphi(x_1, x_2, x_3, t) = \varphi_0 \exp(-ik_1x_1)\exp(-ik_2x_2)\exp(-ik_3x_3)\exp(i\omega t)$$
$$= \varphi_0 \exp[i(\omega t - k_1x_1 - k_2x_2 - k_3x_3)] \tag{9}$$

Note that $k_1x_1 + k_2x_2 + k_3x_3 = \mathbf{k} \cdot \mathbf{x}$, where $\mathbf{x}$ is a position vector defined as

$$\mathbf{x} = x_1\mathbf{e}_1 + x_2\mathbf{e}_2 + x_3\mathbf{e}_3 \tag{10}$$

and $\mathbf{k}$ is the wave-number vector, defined as

$$\mathbf{k} = k_1\mathbf{e}_1 + k_2\mathbf{e}_2 + k_3\mathbf{e}_3 \tag{11}$$

whose magnitude is given by $k^2 = k_1^2 + k_2^2 + k_3^2$. The particular solution of the wave equation is therefore

$$\varphi(\mathbf{x}, t) = \varphi_0 \exp[i(\omega t - \mathbf{k} \cdot \mathbf{x})] \tag{12}$$

8.7.1 P-wave propagation

The scalar potential of P-waves propagating in the direction of the wave-number vector $\mathbf{k}_\alpha$ is

$$\varphi(\mathbf{x}, t) = \varphi_0 \exp[i(\omega t - \mathbf{k}_\alpha \cdot \mathbf{x})] \tag{8.107}$$

The P-wave displacement $\mathbf{u}_P$ is the gradient of φ, and has components

$$\mathbf{u}_P = \nabla \varphi = \mathbf{e}_1 \frac{\partial \varphi}{\partial x_1} + \mathbf{e}_2 \frac{\partial \varphi}{\partial x_2} + \mathbf{e}_3 \frac{\partial \varphi}{\partial x_3} \tag{8.108}$$

This can be written more succinctly using tensor notation:

$$\mathbf{u}_P = \mathbf{e}_n \frac{\partial}{\partial x_n} (\varphi_0 \exp[i(\omega t - k_{\alpha k}x_k)]) = -i\varphi_0(\mathbf{e}_n k_{\alpha n})\exp[i(\omega t - k_{\alpha k}x_k)] \tag{8.109}$$

$$\mathbf{u}_P = -i\varphi_0\mathbf{k}_\alpha \exp[i(\omega t - \mathbf{k}_\alpha \cdot \mathbf{x})] \tag{8.110}$$

Now suppose that the P-wave is propagating in a vertical plane and define the x_1-axis to coincide with the horizontal projection of the direction of propagation. The motions in the P-wave are confined to the x_1–x_3 vertical plane, so there is no displacement in the horizontal x_2-direction and differentiation with respect to x_2 gives zero. The P-wave-number is in this case

$$\mathbf{k}_\alpha = k_{\alpha 1}\mathbf{e}_1 + k_{\alpha 3}\mathbf{e}_3 \tag{8.111}$$

and (8.110) becomes

$$\mathbf{u}_P = -(k_{\alpha 1}\mathbf{e}_1 + k_{\alpha 3}\mathbf{e}_3)i\varphi_0 \exp[i(\omega t - k_{\alpha 1}x_1 - k_{\alpha 3}x_3)] \tag{8.112}$$

The direction of this displacement is the same as that of the ray path or wave-number vector; i.e., the P-wave propagates as an alternation of compressions and rarefactions along the direction of propagation.

8.7.2 S-wave propagation

The vector potential $\boldsymbol{\psi}$ of S-waves propagating in the direction $\mathbf{k}_\beta$ has components

$$\psi_n(\mathbf{x}, t) = \psi_n^0 \exp\left[i(\omega t - \mathbf{k}_\beta \cdot \mathbf{x})\right] \tag{8.113}$$

where the S-wave-number is the vector

$$\mathbf{k}_\beta = k_{\beta 1}\mathbf{e}_1 + k_{\beta 3}\mathbf{e}_3 \tag{8.114}$$

The S-wave displacement $\mathbf{u}_S$ is the curl of $\boldsymbol{\psi}$, and has components

$$\mathbf{u}_S = \nabla \times \boldsymbol{\psi} = \left(\frac{\partial \psi_3}{\partial x_2} - \frac{\partial \psi_2}{\partial x_3}\right)\mathbf{e}_1 + \left(\frac{\partial \psi_1}{\partial x_3} - \frac{\partial \psi_3}{\partial x_1}\right)\mathbf{e}_2 + \left(\frac{\partial \psi_2}{\partial x_1} - \frac{\partial \psi_1}{\partial x_2}\right)\mathbf{e}_3$$
$$\tag{8.115}$$

If we again consider propagation in the x_1–x_3 vertical plane so that differentiation with respect to x_2 gives zero, this equation reduces to

$$\mathbf{u}_S = \left(-\frac{\partial \psi_2}{\partial x_3}\right)\mathbf{e}_1 + \left(\frac{\partial \psi_1}{\partial x_3} - \frac{\partial \psi_3}{\partial x_1}\right)\mathbf{e}_2 + \left(\frac{\partial \psi_2}{\partial x_1}\right)\mathbf{e}_3 \tag{8.116}$$

This can be rearranged as

$$\mathbf{u}_S = \left(-\frac{\partial \psi_2}{\partial x_3}\mathbf{e}_1 + \frac{\partial \psi_2}{\partial x_1}\mathbf{e}_3\right) + \left(\frac{\partial \psi_1}{\partial x_3} - \frac{\partial \psi_3}{\partial x_1}\right)\mathbf{e}_2 \tag{8.117}$$

The second bracketed term on the right describes displacements in the direction of the x_2-axis,

$$\mathbf{u}_{SH} = \left(\frac{\partial \psi_1}{\partial x_3} - \frac{\partial \psi_3}{\partial x_1}\right)\mathbf{e}_2 \tag{8.118}$$

$$\mathbf{u}_{SH} = i(\psi_3^0 k_{\beta 1} - \psi_1^0 k_{\beta 3}) \exp[i(\omega t - \mathbf{k}_\beta \cdot \mathbf{x})] \mathbf{e}_2 \qquad (8.119)$$

The displacements are by definition in the horizontal plane and hence are always normal to the direction of propagation. The horizontal component of a bodily shear wave is known as the SH wave.

The first bracketed term on the right of (8.117) describes a shear wave confined to the vertical x_1–x_3 plane and known as the SV wave. The ψ_2 component of the vector potential in (8.113) is

$$\psi_2 = \psi_2^0 \exp[i(\omega t - k_{\beta 1} x_1 - k_{\beta 3} x_3)] \qquad (8.120)$$

The SV displacement is therefore

$$\mathbf{u}_{SV} = \left(-\frac{\partial \psi_2}{\partial x_3} \mathbf{e}_1 + \frac{\partial \psi_2}{\partial x_1} \mathbf{e}_3\right)$$
$$= (k_{\beta 3} \mathbf{e}_1 - k_{\beta 1} \mathbf{e}_3) i \psi_2^0 \exp[i(\omega t - k_{\beta 1} x_1 - k_{\beta 3} x_3)] \qquad (8.121)$$

The scalar product of the amplitude of the SV displacement vector $\mathbf{u}_{SV}$ and the wave-number $\mathbf{k}_\beta$ is

$$(k_{\beta 3} \mathbf{e}_1 - k_{\beta 1} \mathbf{e}_3) \cdot (k_{\beta 1} \mathbf{e}_1 + k_{\beta 3} \mathbf{e}_3) = 0 \qquad (8.122)$$

This confirms that the SV displacements, like the SH displacements, are normal to the direction of propagation of the S-wave.

These results show that the displacements in the wavefront of a shear wave can be resolved into two orthogonal motions: the SH-component is horizontal and the SV-component is in the vertical plane containing the ray path.

FURTHER READING

Aki, K. and Richards, P. G. (2002). *Quantitative Seismology*, 2nd edn. Sausalito, CA: University Science Books, 704 pp.

Bullen, K. E. (1963). *An Introduction to the Theory of Seismology*, 3rd edn. Cambridge: Cambridge University Press, 381 pp.

Chapman, C. (2004). *Fundamentals of Seismic Wave Propagation*. Cambridge: Cambridge University Press, 172 pp.

Lay, T. and Wallace, T. C. (1995). *Modern Global Seismology*. San Diego, CA: Academic Press, 515 pp.

Shearer, P. M. (2009). *Introduction to Seismology*, 2nd edn. Cambridge: Cambridge University Press, 410 pp.

Udias, A. (2000). *Principles of Seismology*. Cambridge: Cambridge University Press, 490 pp.

Appendix A

Magnetic poles, the dipole field, and current loops

A1. The concept of magnetic poles and Gauss's law

Coulomb carried out experiments with long magnetized needles and showed that their ends exerted forces of attraction and repulsion on the ends of other magnetized needles, similar to the forces between electrical charges. If freely suspended, a magnet aligns in the Earth's own magnetic field so that one end is a north-seeking pole (unfortunately shortened to north pole) and the other a south-seeking pole. Magnetism originates in electric currents, but in some contexts the concept of fictive magnetic poles can be useful. The force between the ends, or poles, of two magnets is proportional to the product of the pole strengths and inversely proportional to the square of the distance r between them. Between two poles of strength p_1 and p_2 the force $\mathbf{F}$ is

$$\mathbf{F} = \frac{\mu_0}{4\pi} \frac{p_1 p_2}{r^2} \mathbf{e}_r \qquad (A1)$$

where μ_0 is the magnetic field constant, or permeability of free space; it is defined to be exactly $4\pi \times 10^{-7} \, \mathrm{N\,A^{-2}}$. The resemblance to Coulomb's law for electrical forces allows us to develop expressions for the magnetic potential and flux. The magnetic field may be defined as the force that acts on a unit magnetic pole. With $p_1 = p$ and $p_2 = 1$, the magnetic field $\mathbf{B}$ of a pole p at distance r is

$$\mathbf{B} = \frac{\mu_0 p}{4\pi r^2} \mathbf{e}_r \qquad (A2)$$

where $\mathbf{e}_r$ is the radial direction. The magnetic potential of a single pole at distance r is therefore

$$W = \int\limits_{r}^{\infty} \mathbf{B} \cdot \mathbf{e}_r \, dr = \frac{\mu_0 p}{4\pi r} \qquad (A3)$$

The flux Φ_m of the magnetic field $\mathbf{B}$ through a surface S surrounding the pole p is

$$\Phi_m = \int_S \mathbf{B} \cdot \mathbf{n} \, dS \tag{A4}$$

where $\mathbf{n}$ is the normal to the surface. Upon inserting the magnetic field $\mathbf{B}$ of the pole from (A2) and defining θ as the angle between $\mathbf{n}$ and the radial direction $\mathbf{e}_r$, the magnetic flux through a surface surrounding the pole p is

$$\Phi_m = \int_S \frac{\mu_0 p}{4\pi r^2} \cos\theta \, dS \tag{A5}$$

Now we make use of the relationship between the solid angle $d\Omega$ subtended at distance r from an inclined surface element dS (Box 1.3), and obtain

$$\Phi_m = \int_{\Omega=0}^{4\pi} \frac{\mu_0 p}{4\pi} \, d\Omega = \mu_0 p \tag{A6}$$

The total pole strength p enclosed by the surface S is therefore given by

$$p = \frac{1}{\mu_0} \Phi_m = \frac{1}{\mu_0} \int_S \mathbf{B} \cdot \mathbf{n} \, dS \tag{A7}$$

Because every magnet has two poles of equal and opposite strength, the sum of all the poles in a volume is zero. The total magnetic flux through any closed surface is therefore also zero. On applying the divergence theorem, we have

$$\Phi_m = \int_S \mathbf{B} \cdot \mathbf{n} \, dS = \int_V \nabla \cdot \mathbf{B} \, dV = 0 \tag{A8}$$

For this to be true for an arbitrary volume

$$\nabla \cdot \mathbf{B} = 0 \tag{A9}$$

This result implies that magnetic monopoles cannot exist. It is known as *Gauss's law* after Carl Friedrich Gauss (1777–1855), who formalized it. The basic magnetic field is that of a dipole.

A2. The magnetic dipole

Two magnetic poles of equal strength but opposite sign, $+p$ and $-p$, are a distance d apart (Fig. A1). The geometry has rotational symmetry about the line AB joining the poles, the magnetic axis. The radius of length r from the point M midway between the poles to the point P, where the magnetic potential is to be determined, makes an angle θ with the magnetic axis. Let the distance of P from the positive pole be $r^{(+)}$ and the distance from the negative pole be $r^{(-)}$. Following (A3), the potential of the positive pole at P is

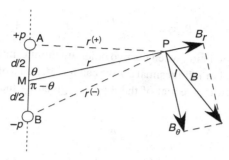

Fig. A1. The geometry for calculation of the magnetic potential and the radial and azimuthal fields of a pair of opposite and equal magnetic poles. In the limit, as the separation of the poles tends to zero, the potential and fields are those of a magnetic dipole.

$$W^{(+)} = \frac{\mu_0}{4\pi} \frac{p}{r^{(+)}} \tag{A10}$$

On applying the reciprocal-distance definition of the Legendre polynomials (Section 1.12, Fig. 1.11) to the triangle AMP, this potential expands to

$$W^{(+)} = \frac{\mu_0 p}{4\pi r} \left(1 + \sum_{n=1}^{\infty} \left(\frac{d}{2r} \right)^n P_n(\cos\theta) \right) \tag{A11}$$

Similarly, for the negative pole, the relations of the sides in the triangle BMP give

$$W^{(-)} = -\frac{\mu_0}{4\pi} \frac{p}{r^{(-)}} = -\frac{\mu_0 p}{4\pi r} \left(1 + \sum_{n=1}^{\infty} \left(\frac{d}{2r} \right)^n P_n(\cos(\pi - \theta)) \right) \tag{A12}$$

The combined potential of both magnetic poles at the point P is

$$W = \frac{\mu_0 p}{4\pi r} \left\{ \sum_{n=1}^{\infty} \left(\frac{d}{2r} \right)^n [P_n(\cos\theta) - P_n(-\cos\theta)] \right\} \tag{A13}$$

From Rodrigues' formula (Section 1.14) we find that

$$P_n(-x) = \frac{1}{2^n n!} (-1)^n \frac{d^n}{dx^n} (x^2 - 1)^n = (-1)^n P_n(x) \tag{A14}$$

The potential of the magnetic pole-pair is thus

$$W = \frac{\mu_0 p}{4\pi r} \sum_{n=1}^{\infty} \left(\frac{d}{2r} \right)^n (P_n(\cos\theta) - (-1)^n P_n(\cos\theta)) \tag{A15}$$

Each successive term is smaller than the previous term by the ratio $d/(2r)$. The first terms are

$$W = \frac{\mu_0}{4\pi} \frac{pd}{r^2} P_1(\cos\theta) + \frac{\mu_0}{4\pi} \frac{pd}{r^2} \left(\frac{d}{2r}\right)^2 P_3(\cos\theta) + \cdots \tag{A16}$$

A dipole is the constellation when the two poles are infinitesimally close to each other, so that $d \ll r$. For infinitesimal d/r we can ignore terms of higher than first order, so the magnetic potential of the dipole is given by the first term in the equation, which we can write

$$W = \frac{\mu_0}{4\pi} \frac{m\cos\theta}{r^2} \tag{A17}$$

The quantity $m = pd$ is called the *magnetic moment* of the dipole, for the following reason. A dipole of length d, whose axis makes an angle θ with a uniform magnetic field **B**, experiences a force $+p\mathbf{B}$ on one pole and an opposite force $-p\mathbf{B}$ on the other pole. The perpendicular distance between the lines of action of these forces is $d \sin\theta$, so the field exerts a torque $\boldsymbol{\tau}$ of magnitude $pdB\sin\theta$ in the direction normal to both the field and the dipole.

$$\tau = pdB\sin\theta = mB\sin\theta \tag{A18}$$

$$\boldsymbol{\tau} = \mathbf{m} \times \mathbf{B} \tag{A19}$$

The magnetic moment **m** of the dipole is a vector oriented along the dipole axis from the negative to the positive pole.

A3. The Lorentz force

When an electrical charge q moves with velocity **v** through a magnetic field **B**, there arises a force **F** that is normal both to the field and to the direction of motion (Fig. A2(a)). This is the *Lorentz* force, which serves to define the unit of magnetic field,

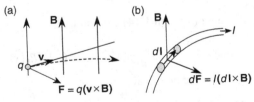

Fig. A2. (a) The Lorentz force **F** on a charged particle moving with velocity **v** in a magnetic field **B** acts normal to both the velocity and the field, resulting in a curved trajectory (dashed line). (b) The Biot–Savart law gives the increment of force $d\mathbf{F}$ experienced by a short conductor of length $d\mathbf{l}$ carrying a current I in a magnetic field **B**. After Lowrie (2007).

$$\mathbf{F} = q(\mathbf{v} \times \mathbf{B}) \qquad (A20)$$

With force measured in newtons (N), charge in coulombs (C), velocity in meters per second (m s^{-1}), and electric current in amperes (A = C s^{-1}), the unit of magnetic field is the tesla, which has the dimensions N A^{-1} m^{-1}.

Imagine the moving charge to be confined to move along a conductor of length dl and cross-section A (Fig. A2(b)). Let the number of charges per unit volume be N. The total charge inside the element of length dl is then $NAq\,dl$ and the Lorentz force acting on the element $d\mathbf{l}$ is

$$d\mathbf{F} = NAq\,dl(\mathbf{v} \times \mathbf{B}) \qquad (A21)$$

The current $\mathbf{v}$ and the element $d\mathbf{l}$ of the conductor have the same direction, so we can write

$$d\mathbf{F} = NAqv(d\mathbf{l} \times \mathbf{B}) \qquad (A22)$$

The electric current I along the conductor is the total charge that crosses a surface A per second; it is equal to $NAqv$. The force experienced by the element $d\mathbf{l}$ of a conductor carrying a current I in a magnetic field $\mathbf{B}$ is therefore

$$d\mathbf{F} = I(d\mathbf{l} \times \mathbf{B}) \qquad (A23)$$

A4. Torque on a current loop in a magnetic field

Using (A23), we can compute the force acting on each side of a small rectangular loop PQRS, which carries an electric current I in a magnetic field $\mathbf{B}$ (Fig. A3(a)). Let the lengths of the sides of the loop be a and b, respectively, and let the x-axis be parallel to the sides of length a. The area A of the loop is equal to ab; $\mathbf{n}$ is the direction normal to the plane of the loop. The magnetic field $\mathbf{B}$ acts

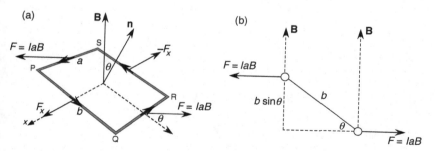

Fig. A3. (a) Forces on the sides a and b of a rectangular coil whose plane is inclined at angle θ to a magnetic field $\mathbf{B}$. (b) Cross-section showing how the equal and opposite, but not collinear, forces produce a torque on the coil. After Lowrie (2007).

normal to the x-axis, making an angle θ with the direction **n**. A force F_x equal to $IbB \cos \theta$ acts on the side PQ in the direction of $+x$, and an equal and opposite force F_x acts on the side RS in the direction of $-x$; these forces are collinear and cancel each other out. Forces equal to IaB act in opposite directions on the sides QR and SP and the perpendicular distance between their lines of action is $b \sin \theta$ (Fig. A3(b)), so the magnitude of the torque τ experienced by the current loop is

$$\tau = IaBb \sin \theta = IAB \sin \theta = mB \sin \theta \tag{A24}$$

$$\boldsymbol{\tau} = \mathbf{m} \times \mathbf{B} \tag{A25}$$

The quantity $\mathbf{m} = IA\mathbf{n}$ is a vector normal to the plane of the current loop. Comparison with (A19) shows that it corresponds to the magnetic moment of the current loop. At distances much greater than the dimensions of the loop, the magnetic field is that of a dipole at the center of the loop. Consequently, magnetic behavior is more correctly explained by replacing fictive magnetic dipoles by current loops. This is true even at atomic dimensions; circulating (and spinning) electrical charges impart magnetic moments to atoms. The definition of **m** in terms of a current-carrying loop shows that magnetic moment has the dimensions of current times area, or ampere meter^2 (A m^2).

Appendix B

Maxwell's equations of electromagnetism

In the early nineteenth century, experimental observations of electrical and magnetic behavior led to the establishment of fundamental physical laws governing electricity and magnetism. In 1873 the Scottish scientist James Clerk Maxwell synthesized all known empirical laws of electricity and magnetism into a set of equations that describe electromagnetic phenomena. They embody in succinct form the empirical laws of Coulomb, Ampère, Gauss, and Faraday.

1. Coulomb's law

Charles Augustin de Coulomb (1736–1806) discovered experimentally that the force $\mathbf{F}$ between two electrical charges Q_1 and Q_2 is proportional to the product of the charges and inversely proportional to the square of the distance r between them. Let $\mathbf{e}_r$ be the unit vector from Q_1 to Q_2. In the international system (SI) of units Coulomb's law is

$$\mathbf{F} = \frac{Q_1 Q_2}{4\pi\varepsilon_0 r^2}\mathbf{e}_r \tag{B1}$$

In this equation ε_0 is the *electric field constant,* or the *permittivity* of free space; it is equal to $8.854\,187\,817 \times 10^{-12}$ C^2 N^{-1} m^{-2}. If both charges are positive or negative, the force between them is repulsive; if the charges have opposite sign, the force is attractive.

The electric field $\mathbf{E}$ is defined as the force that acts on a unit positive electrical charge. If we let $Q_1 = Q$ and $Q_2 = 1$, the electric field of the charge Q at distance r is

$$\mathbf{E} = \frac{Q}{4\pi\varepsilon_0 r^2}\mathbf{e}_r \tag{B2}$$

If the charge Q is positive, the field acts outwards, in the direction of increasing r. The electric potential at distance r is

$$U = \int_r^\infty \mathbf{E} \cdot \mathbf{e}_r \, dr = \frac{Q}{4\pi\varepsilon_0 r} \tag{B3}$$

The flux Φ of the electric field $\mathbf{E}$ through a surface S surrounding the charge Q is

$$\Phi = \int_S \mathbf{E} \cdot \mathbf{n} \, dS = \int_S \frac{Q}{4\pi\varepsilon_0 r^2} (\mathbf{e}_r \cdot \mathbf{n}) \, dS \tag{B4}$$

where $\mathbf{n}$ is the unit vector normal to the surface element dS. If θ is the angle between $\mathbf{n}$ and the radial direction $\mathbf{e}_r$, the scalar product of the unit vectors equals $\cos\theta$, therefore

$$\Phi = \int_S \frac{Q}{4\pi\varepsilon_0 r^2} \cos\theta \, dS \tag{B5}$$

We can use the definition of a solid angle (Box 1.3) to change the surface integral to an integral over a solid angle around the charge Q:

$$\Phi = \int_S \frac{Q}{4\pi\varepsilon_0} \frac{\cos\theta}{r^2} \, dS = \int_0^{4\pi} \frac{Q}{4\pi\varepsilon_0} \, d\Omega = \frac{Q}{\varepsilon_0} \tag{B6}$$

$$Q = \varepsilon_0 \Phi = \varepsilon_0 \int_S \mathbf{E} \cdot \mathbf{n} \, dS \tag{B7}$$

If the electrical charge Q is distributed throughout a volume V with charge density ρ,

$$Q = \int_V \rho \, dV \tag{B8}$$

We can apply Gauss's divergence theorem to the right-hand side of (B7), which becomes

$$\int_V \rho \, dV = \varepsilon_0 \int_S \mathbf{E} \cdot \mathbf{n} \, dS = \varepsilon_0 \int_V \nabla \cdot \mathbf{E} \, dV \tag{B9}$$

$$\int_V (\rho - \varepsilon_0 \nabla \cdot \mathbf{E}) \, dV = 0 \tag{B10}$$

The volume V is arbitrary, so the integrand must always be zero. This gives Coulomb's law for the field of free electrical charges with density distribution ρ:

$$\nabla \cdot \mathbf{E} = \frac{\rho}{\varepsilon_0} \tag{B11}$$

1.1. The effect of bound charges

In some materials, called *dielectrics*, electrical charges are not free, but are bound to atoms in fixed locations. An applied electric field can cause the bound charges to shift position (e.g., from one side of an atom to the other), with positive and negative charges displaced in opposite directions. This results in an electric polarization **P**. A charge Q_D accumulates on an arbitrary surface S within a homogeneous dielectric material, equivalent to

$$Q_D = \int_S \mathbf{P} \cdot \mathbf{n} \, dS \tag{B12}$$

The total charge Q_T carried by a polarizable material is the sum of the free charge Q and the bound surface charge Q_D:

$$Q_T = Q + Q_D \tag{B13}$$

$$\int_V \rho_T \, dV = \varepsilon_0 \int_S \mathbf{E} \cdot \mathbf{n} \, dS + \int_S \mathbf{P} \cdot \mathbf{n} \, dS \tag{B14}$$

Gauss's theorem allows us to convert the surface integrals into volume integrals:

$$\int_V \rho_T \, dV = \varepsilon_0 \int_V \nabla \cdot \mathbf{E} \, dV + \int_V \nabla \cdot \mathbf{P} \, dV \tag{B15}$$

It follows that

$$\nabla \cdot (\varepsilon_0 \mathbf{E} + \mathbf{P}) = \rho_T \tag{B16}$$

The electric displacement vector **D** is defined by

$$\mathbf{D} = \varepsilon_0 \mathbf{E} + \mathbf{P} \tag{B17}$$

Coulomb's law for a material that can be polarized electrically is therefore

$$\nabla \cdot \mathbf{D} = \rho_T \tag{B18}$$

In a homogeneous dielectric material the electric polarization **P** is proportional to the electric field **E**. In SI usage the proportionality constant is written as the product of the permittivity ε_0 and the *electric susceptibility* χ. Thus

$$\mathbf{P} = \chi \varepsilon_0 \mathbf{E} \tag{B19}$$

$$\mathbf{D} = \varepsilon_0 \mathbf{E} + \chi \varepsilon_0 \mathbf{E} \tag{B20}$$

$$\mathbf{D} = (1 + \chi) \varepsilon_0 \mathbf{E} = \varepsilon \varepsilon_0 \mathbf{E} \tag{B21}$$

The dimensionless quantity ε is the *relative permittivity*, or *dielectric constant*, of the material. In a material that cannot be polarized $\varepsilon = 1$ and

$$\mathbf{D} = \varepsilon_0 \mathbf{E} \tag{B22}$$

In this case, if the density of free charges is ρ,

$$\nabla \cdot \mathbf{D} = \rho \tag{B23}$$

2. Ampère's law

Ampère's law describes magnetic fields produced by electric currents. Experiments begun in 1820 by André-Marie Ampère (1775–1836) and Hans Christian Ørsted (1777–1851) showed that an electric current produces a magnetic field. Ampère's experiments on a long, straight, electrical conductor showed that the magnetic field is in the plane normal to the conductor, and the field direction obeys a right-hand rule with respect to the current (i.e., the directions of current and field are indicated by the thumb and fingers, respectively). For example, the field lines around a long straight conductor are concentric circles (Fig. B1(a)). The strength of the magnetic field outside the conductor is proportional to the current I in the conductor and inversely proportional to the distance r from the conductor:

$$B \propto \frac{I}{r} \tag{B24}$$

In general, if $d\mathbf{l}$ is an element of the closed path L around a conductor carrying a current I in a magnetic field $\mathbf{B}$, Ampère's law is

$$\oint_L \mathbf{B} \cdot d\mathbf{l} = \mu_0 I \tag{B25}$$

The magnetic field constant μ_0 ensures compatibility between the units of electric current and magnetic field. The integration can also be applied to a

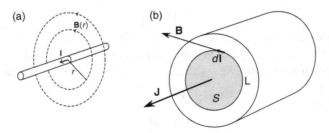

Fig. B1. (a) The lines of magnetic field **B** around a long straight conductor carrying an electric current I are concentric circles. (b) For a path inside an electrical conductor only the fraction of the current enclosed by the path causes the magnetic field **B** along the path.

path L inside an electrical conductor, at right angles to the flow of current (Fig. B1(b)). In this case, not all the current is enclosed by the loop, and only the fraction of the current passing through the loop causes the magnetic field **B**. If **J** is the electric current density (i.e., the current per unit cross-sectional area normal to the flow), the amount of current enclosed by the loop is

$$I = \int_S \mathbf{J} \cdot \mathbf{n} \, dS \tag{B26}$$

Equating this with (B25) gives

$$\oint_L \mathbf{B} \cdot d\mathbf{l} = \mu_0 \int_S \mathbf{J} \cdot \mathbf{n} \, dS \tag{B27}$$

We now use Stokes' theorem to convert the left-hand side into a surface integral:

$$\int_S \nabla \times \mathbf{B} \cdot \mathbf{n} \, dS = \mu_0 \int_S \mathbf{J} \cdot \mathbf{n} \, dS \tag{B28}$$

This must be true for any surface intersecting the current, thus

$$\nabla \times \mathbf{B} = \mu_0 \mathbf{J} \tag{B29}$$

This is Ampère's law for the magnetic field produced by an electric current in a conductor.

The current density **J** is proportional to the electric field **E**. This follows from *Ohm's law*, which relates the current (I) and voltage (V) to the resistance (R) of a circuit:

$$V = IR \tag{B30}$$

The electric field E is the voltage per unit distance along a circuit. In a straight conductor of length L and cross-sectional area A the voltage V equals EL and the current I equals JA. The resistance R of a conductor is proportional to its length L and inversely proportional to its cross-sectional area A. The constant of proportionality is the resistivity; its inverse is the conductivity, σ. Consequently $R = (1/\sigma)L/A$ and substitution into Ohm's law gives

$$(EL) = (JA)\left(\frac{L}{\sigma A}\right) \tag{B31}$$

After simplifying, we get Ohm's law in vector form:

$$\mathbf{J} = \sigma\mathbf{E} \tag{B32}$$

By combining this with (B29), we get an alternative form of Ampère's law:

$$\nabla \times \mathbf{B} = \mu_0 \sigma \mathbf{E} \tag{B33}$$

This law applies to the magnetic effect produced by a current of free electrical charges. However, bound electrical charges can also result in an electric current and produce a magnetic field.

2.1. The effect of displacement currents

In a dielectric material, the electrical charges are bound to atoms, but a time-dependent change in their positions is equivalent to a *displacement current* I_D. The total electric current I_T is the sum of the current I passing through the material and the displacement current I_D. Differentiating (B13) gives

$$\frac{\partial}{\partial t} Q_T = \frac{\partial}{\partial t} Q + \frac{\partial}{\partial t} Q_D \tag{B34}$$

Using (B26) and writing the volume density of the bound charges as ρ_D,

$$\int_S \mathbf{J}_T \cdot \mathbf{n}\, dS = \int_S \mathbf{J} \cdot \mathbf{n}\, dS + \frac{\partial}{\partial t} \int_V \rho_D\, dV \tag{B35}$$

Applying Gauss's theorem to the first two terms and using the result of (B18) gives

$$\int_V (\nabla \cdot \mathbf{J}_T) dV = \int_V (\nabla \cdot \mathbf{J}) dV + \frac{\partial}{\partial t} \int_V (\nabla \cdot \mathbf{D}) dV \tag{B36}$$

The total current density, combining the free charges and bound charges, is

$$\mathbf{J}_T = \mathbf{J} + \frac{\partial \mathbf{D}}{\partial t} \tag{B37}$$

Using the total current density in Ampère's equation, we get

$$\nabla \times \mathbf{B} = \mu_0 \mathbf{J} + \mu_0 \frac{\partial \mathbf{D}}{\partial t} \tag{B38}$$

Finally, using Ohm's law (B32) and the relation between the electric displacement vector and the electric field (B22), Ampère's law for a non-polarizable medium is

$$\nabla \times \mathbf{B} = \mu_0 \sigma \mathbf{E} + \mu_0 \varepsilon_0 \frac{\partial \mathbf{E}}{\partial t} \tag{B39}$$

3. Gauss's law for magnetism

Early experimenters concluded that, unlike electrical charges, magnetic monopoles did not exist. Division of a magnet into smaller pieces always left a number of magnets with two poles. All magnetic fields originate from electric currents,

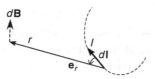

Fig. B2. At distance r in a direction $\mathbf{e}_r$ from a short conductor of length $d\mathbf{l}$ carrying a current I the magnetic field $d\mathbf{B}$ is normal to both $d\mathbf{l}$ and $\mathbf{e}_r$.

whether at macroscopic or at microscopic (atomic) level. Ampère's investigations were extended by his contemporaries, Jean-Baptiste Biot (1774–1862) and Félix Savart (1791–1841). Their empirical studies of the forces between straight conductors carrying electric currents showed that the magnetic field $d\mathbf{B}$ at a distance r from a short conductor of length $d\mathbf{l}$ carrying a current I is given by

$$d\mathbf{B} = \frac{\mu_0}{4\pi r^2} I(d\mathbf{l} \times \mathbf{e}_r) \tag{B40}$$

The unit vector (direction) $\mathbf{e}_r$ is from the current element to the point of observation (Fig. B2). The total field of a current circuit at the point of observation P is found by integrating (B40) around the circuit, which necessarily depends on the geometry of the circuit.

It follows that the magnetic field is divergence-free. Taking the divergence of (B40) gives

$$\nabla \cdot d\mathbf{B} = \frac{\mu_0 I}{4\pi} \nabla \cdot \left(\frac{d\mathbf{l} \times \mathbf{e}_r}{r^2}\right) \tag{B41}$$

The length of the current element $d\mathbf{l}$ is constant with respect to the differentiation. The order of the differentiation can be changed, changing sign accordingly, which gives

$$\nabla \cdot d\mathbf{B} = -\frac{\mu_0 I}{4\pi} d\mathbf{l} \cdot \left(\nabla \times \frac{\mathbf{e}_r}{r^2}\right) \tag{B42}$$

The function of r to be differentiated is recognizable as

$$\frac{\mathbf{e}_r}{r^2} = -\nabla\left(\frac{1}{r}\right) \tag{B43}$$

Substitution into (B42) leads to the curl of a gradient, which is always zero (see (1.32)):

$$\nabla \cdot d\mathbf{B} = -\frac{\mu_0 I}{4\pi} d\mathbf{l} \cdot \left(\nabla \times \nabla\left(\frac{1}{r}\right)\right) = 0 \tag{B44}$$

If this is true for every contribution $d\mathbf{B}$ to the field, it must be true for the entire field. This yields Gauss's law for magnetism:

$$\nabla \cdot \mathbf{B} = 0$$

Let V be an arbitrary volume enclosed by a surface S in a magnetic field $\mathbf{B}$. The net flux of the magnetic field through the surface is obtained using Gauss's divergence theorem (Section 1.6):

$$\int_S (\mathbf{B} \cdot \mathbf{n}) dS = \int_V (\nabla \cdot \mathbf{B}) dV = 0 \tag{B45}$$

The net flux of the magnetic field through the surface is always zero; the number of field lines entering the surface is the same as the number leaving the surface. Hence magnetic field lines always form complete loops; they do not begin or end on "charges" as the electric field does. This implies that magnetic monopoles do not exist. The elementary magnetic field is that of a dipole.

3.1. The magnetic field inside a magnetizable material

Just as bound charges affect the electric field inside a dielectric, the magnetic field inside a magnetically polarizable material is modified by the internal electric currents in the material. The atoms in crystalline materials occupy fixed positions in a regular lattice structure and their atomic magnetic moments can be partially aligned by a magnetic field. The net magnetic moment per unit volume of the material is its magnetization, $\mathbf{M}$. Consider a small volume element with sides Δx, Δy, and Δz at the point (x, y, z) in a magnetizable material (Fig. B3). A current I_1 flows around the small loop with sides Δy and Δz, causing a magnetization component M_x in the x-direction. The magnetic moment of a current loop is the product of its area and the current in the loop (Appendix A4):

$$m_x = M_x \Delta V = M_x \Delta x \Delta y \Delta z = I_1 \Delta y \Delta z \tag{B46}$$

$$I_1 = M_x \Delta x \tag{B47}$$

The magnetization is not necessarily uniform, so in the adjacent loop in the y-direction it may equal $(M_x + \Delta M_x)$ with a circulation current I_2, where

$$I_2 = (M_x + \Delta M_x)\Delta x = \left(M_x + \frac{\partial M_x}{\partial y} \Delta y \right) \Delta x \tag{B48}$$

The net current at the interface between the loops is in the z-direction. Its magnitude is the difference between I_1 and I_2:

$$I_z = I_1 - I_2 = -\frac{\partial M_x}{\partial y} \Delta y \Delta x \tag{B49}$$

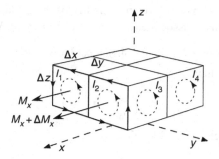

Fig. B3. Production of magnetization components M_x and $M_x + \Delta M_x$ in the x-direction from currents I_1 and I_2 in adjacent small loops in the y–z plane within a magnetizable material.

If **J** is the current density in the material, the z-component of current must equal $J_z \, \Delta x \, \Delta y$. The x-component of magnetization thus makes a contribution to the current density in the z-direction equal to

$$J_z = -\frac{\partial M_x}{\partial y} \tag{B50}$$

A similar argument can be applied to the current loops in the x–z plane, which carry currents I_3 and I_4, respectively, causing magnetization components M_y and $(M_y + \Delta M_y)$. Taking into account the sense of the currents around the small loops, the net current in the z-direction from these loops is

$$I_z = I_4 - I_3 = \frac{\partial M_y}{\partial x} \Delta z \, \Delta x \tag{B51}$$

The corresponding contribution to the current density in the z-direction is

$$J_z = \frac{\partial M_y}{\partial x} \tag{B52}$$

The net z-component of the current density is found by combining (B50) and (B52):

$$J_z = \frac{\partial M_y}{\partial x} - \frac{\partial M_x}{\partial y} = (\nabla \times \mathbf{M})_z \tag{B53}$$

By treating the current circulation in other pairs of the reference planes, the other components of **J** can be obtained. The current density $\mathbf{J}_m$ associated with the magnetization **M** is therefore

$$\mathbf{J}_m = \nabla \times \mathbf{M} \tag{B54}$$

Inside a magnetizable material we must modify Ampère's law (B29) by adding the extra current density associated with the magnetization. We then get

$$\nabla \times \mathbf{B} = \mu_0 (\mathbf{J} + \mathbf{J}_m) = \mu_0 (\mathbf{J} + \nabla \times \mathbf{M}) \qquad (B55)$$

On rearranging, we have

$$\nabla \times \left(\frac{\mathbf{B}}{\mu_0} - \mathbf{M} \right) = \mathbf{J} \qquad (B56)$$

Let an auxiliary vector **H** be defined as

$$\mathbf{H} = \frac{\mathbf{B}}{\mu_0} - \mathbf{M} \qquad (B57)$$

$$\mathbf{B} = \mu_0 (\mathbf{H} + \mathbf{M}) \qquad (B58)$$

H has the same dimensions (A m^{-1}) as magnetization. Historically it has been called the magnetizing field, despite having the wrong dimensions. Inside an isotropic, non-ferromagnetic material the magnetization **M** is proportional to **H**:

$$\mathbf{M} = \chi \mathbf{H} \qquad (B59)$$

The constant of proportionality is the magnetic susceptibility, χ, which is a dimensionless property of the material. The relationship between **B** and **H** is thus

$$\mathbf{B} = \mu_0 \mathbf{H} (1 + \chi) = \mu \mu_0 \mathbf{H} \qquad (B60)$$

The quantity $\mu = 1 + \chi$ is the magnetic permeability of the material. In free space and in materials that cannot acquire a magnetization the susceptibility is zero and the permeability $\mu = 1$, so

$$\mathbf{B} = \mu_0 \mathbf{H} \qquad (B61)$$

4. Faraday's law

In 1831 an English scientist, Michael Faraday (1791–1867), demonstrated that a change in the magnetic flux Φ_m through a coil induced in the coil an electric voltage V proportional to the rate of change of the flux. The direction of the induced voltage was shown by Heinrich Lenz (1804–1865) to oppose the change in flux through the coil. Thus

$$V = -\frac{\partial}{\partial t} \Phi_m \qquad (B62)$$

The flux of the magnetic field through a coil with surface area S is

$$\Phi_m = \int_S \mathbf{B} \cdot \mathbf{n} \, dS \qquad (B63)$$

If **E** is the electric field induced in the coil, and $d\mathbf{l}$ is an element of the wire in the coil, the voltage induced in a path of length L (e.g., a circumference of the coil) is

$$V = \int_L \mathbf{E} \cdot d\mathbf{l}$$

(B64)

With the aid of Stokes' theorem the linear integral around the closed path L can be converted into a surface integral over the area S enclosed by L:

$$V = \int_S \nabla \times \mathbf{E} \cdot \mathbf{n}\, dS$$

(B65)

Combining (B62), (B63), and (B65) gives

$$\int_S \nabla \times \mathbf{E} \cdot \mathbf{n}\, dS = -\frac{\partial}{\partial t} \int_S \mathbf{B} \cdot \mathbf{n}\, dS$$

(B66)

It follows that

$$\nabla \times \mathbf{E} = -\frac{\partial \mathbf{B}}{\partial t}$$

(B67)

This is Faraday's law describing the generation of an electric field from a changing magnetic field.

References

Cain, J. C., Wang, Z., Schmitz, D. R., and Meyer, J. (1989). The geomagnetic spectrum for 1980 and core–crustal separation. *Geophys. J.*, **97**, 443–447.

Creer, K. M., Georgi, D. T., and Lowrie, W. (1973). On the representation of the Quaternary and Late Tertiary geomagnetic fields in terms of dipoles and quadrupoles. *Geophys. J. R. Astron. Soc.*, **33**, 323–345.

Dyson, F. and Furner, H. (1923). The earth's magnetic potential. *Mon. Not. R. Astron. Soc. Geophys. Suppl.*, **1**, 76–88.

Dziewonski, A. M. and Anderson, D. L. (1981). Preliminary Reference Earth Model (PREM). *Phys. Earth Planet. Inter.*, **25**, 297–356.

Finlay, C. C., Maus, S., Beggan, C. D. *et al.* (2010). International Geomagnetic Reference Field: The Eleventh Generation. *Geophys. J. Int.*, **183**, 1216–1230.

Groten, E. (2004). Fundamental parameters and current (2004) best estimates of the parameters of common relevance to astronomy, geodesy, and geodynamics. *J. Geodesy*, **77**, 724–731.

Hasterok, D. P. (2010). *Thermal State of Continental and Oceanic Lithosphere*, Ph.D. thesis, University of Utah, Salt Lake City, USA.

Lowes, F. J. (1966). Mean square values on sphere of spherical harmonic vector fields. *J. Geophys. Res.*, **71**, 2179.

(1974). Spatial power spectrum of the main geomagnetic field, and extrapolation to the core. *Geophys. J. R. Astron. Soc.*, **36**, 717–730.

(1994). The geomagnetic eccentric dipole: facts and fallacies. *Geophys. J. Int.*, **118**, 671–679.

Lowrie, W. (2007). *Fundamentals of Geophysics*, 2nd edn. Cambridge: Cambridge University Press, 381 pp.

McCarthy, D. D. and Petit, G. (2004). *IERS Conventions (2003)*, IERS Technical Note No. 32. Frankfurt am Main: Verlag des Bundesamtes für Kartographie und Geodäsie, 127 pp.

Schmidt, A. (1934). Der magnetische Mittelpunkt der Erde und seine Bedeutung. *Gerlands Beiträge zur Geophysik*, **41**, 346–358.

Stacey, F. D. (1992). *Physics of the Earth*, 3rd edn. Brisbane: Brookfield Press, 513 pp.
(2007). Core properties, physical, in *Encyclopedia of Geomagnetism and Paleomagnetism*, ed. D. Gubbins and E. Herrero-Bervera. Dordrecht: Springer, pp. 91–94.

Stacey, F. D. and Anderson, O. L. (2001). Electrical and thermal conductivities of Fe–Ni–Si alloy under core conditions. *Phys. Earth Planet. Inter.*, **124**, 153–162.

Stacey, F. D. and Davis, P. M. (2008). *Physics of the Earth*, 4th edn. Cambridge: Cambridge University Press, 532 pp.

Vosteen, H.-D. and Schellschmidt, R. (2003). Influence of temperature on thermal conductivity, thermal capacity and thermal diffusivity for different types of rock. *Phys. Chem. Earth*, **28**, 499–509.

Index

278

Printed in the United States
By Bookmasters